PRAISE FOR
THE HEALTHY VEGETABLE GARDEN

A must read for anyone who wants to know how to grow their own zero–food miles, pesticide-free veg, while treading gently upon our planet.

Dave Goulson, author of *The Garden Jungle* and *Gardening for Bumblebees*

COVID has led to a renewed appreciation of nature for many people, and an interest in joining CSAs or growing their own produce. So this book could not be better timed, and given Sally's lifetime experience of organic gardening, it's bound to inspire all those who want to 'grow back better'.

Helen Browning, chief executive, Soil Association

If you want to grow your own food while helping nature to thrive, this fascinating book is for you. I learnt a lot.

Alex Mitchell, author of *Crops in Tight Spots*

An invaluable addition to the gardening literature, Sally Morgan's new book does a superb job of presenting a holistic, ecological approach to soil health as well as pest and disease management. Whereas many books may simply say, 'Use X product to fight X issue', Morgan describes in thorough detail what the gardener can do – from building habitats to improving soil conditions – to ensure that issue never has a chance to take hold. You will turn to this book time and again as a smart, non-dogmatic guide to producing consistently healthy crops and, in effect, a healthier environment.

Jesse Frost, author of *The Living Soil Handbook*

There are so many reasons to grow a healthy food garden, and expert gardener Sally Morgan shows you how. In this comprehensive guide, you will dig deep into the world of soil, natural pest control and companion plants, along with a range of options for your natural landscape. This clearly written book sets up a strong foundation for new gardeners while also offering advice for those with more experience, who seek tools for building and maintaining a healthy food garden from start to finish.

Ellen Ecker Ogden, author of *The New Heirloom Garden*

In *The Healthy Vegetable Garden,* Sally Morgan shares a wealth of information about growing healthy soil as the foundation for exceptional gardens which produce nutritional food with minimal pests. Her simple informative style empowers the reader to put these concepts into grounded action. She has created a fantastic resource for the success of the next generation of gardeners.

Katrina Blair, author of *The Wild Wisdom of Weeds*

One thing I found refreshing about Sally Morgan's book is that she doesn't trade in the myth of the 'perfectly balanced organic soil' which, of course, is not to be found in this imperfect, constantly evolving world. She doesn't aspire to a weed-free or pest-free garden, but rather one where problems are kept manageable without resorting to extreme or disruptive remedies. She cites a slew of strategies one might use to correct course: biocontrols, barriers repellents, and so on. However, she keeps returning to the basic premise that by fostering a diverse and resilient habitat, the soil and garden ecosystem can do a great deal to regenerate and maintain itself, making external fixes unnecessary.

Will Bonsall, author of *Will Bonsall's Essential Guide to Radical, Self-Reliant Gardening*

The Resilient Garden and Allotment Handbook

Also by Sally Morgan

Living on One Acre or Less
(Green Books, 2016)

The Climate Change Garden
with Kim Stoddart
(Cool Springs Press, 2022)

The Resilient Garden and Allotment Handbook

Enrich your soil,
manage pests and diseases
and boost biodiversity
without toxic chemicals and
synthetic fertilisers

SALLY MORGAN

Chelsea Green Publishing
London, UK
White River Junction, Vermont

Copyright © 2021, 2024 by Sally Morgan.
All rights reserved.

Previously published as *The Healthy Vegetable Garden*, now updated and revised.

No part of this book may be transmitted or reproduced in any form by any means without permission in writing from the publisher.

Developmental Editor: Muna Reyal
Project Manager: Rebecca Springer
Copy Editor: Caroline West
Proofreader: Nancy A. Crompton
Indexer: WordCo Indexing Services, Inc.
Designer: Melissa Jacobson
Page Layout: Abrah Griggs

Printed in the United Kingdom.
First printing February 2024.
10 9 8 7 6 5 4 3 2 1   24 25 26 27 28

Our Commitment to Green Publishing

Chelsea Green sees publishing as a tool for cultural change and ecological stewardship. We strive to align our book manufacturing practices with our editorial mission and to reduce the impact of our business enterprise in the environment. We print our books using vegetable-based inks whenever possible. This book may cost slightly more because it was printed on paper from responsibly managed forests, and we hope you'll agree that it's worth it. *The Resilient Garden and Allotment Handbook* was printed on paper supplied by T.J. Books that is certified by the Forest Stewardship Council.

ISBN 978-1-915294-56-2 (paperback)
ISBN 978-1-915294-57-9 (ebook)

Chelsea Green Publishing
London, UK
White River Junction, Vermont, USA

www.chelseagreen.co.uk

Contents

Introduction — 1

PART 1 Building a healthy soil
1 Soil basics — 7
2 Regenerating your soil — 21

PART 2 Pests and predators
3 Understanding pests and diseases — 45
4 Natural predators — 65

PART 3 Plant diversity
5 Getting the planting right — 81
6 Choosing the right plant — 107

PART 4 Boosting defences
7 Biocontrol — 117
8 Plant defences — 131
9 Barriers, lures, traps and sprays — 143

PART 5 Managing water
10 Be water-wise — 153

A look to the future — 169

Index — *175*

Introduction

Resilience. It's a word that is used a lot in our everyday lives, describing the ability to withstand challenges, bounce back from setbacks, and emerge stronger, more adaptable, and better prepared for future challenges. As our climate undergoes profound shifts and our ecosystems face unprecedented challenges, the need for resilient gardening practices has never been more urgent.

In recent years, gardeners have been faced with all manner of challenges, from strong winds and heavy rainfall, to extreme heat and cold and weeks of drought, as climate change begins to make itself felt. We need to prepare and change the way we garden to cope with the future conditions. Our gardens and allotments need to be adaptable, to be able to recover from adverse events, and continue to thrive in the face of disruptions. *The Resilient Garden and Allotment Handbook* aims to be more than a mere guide, rather it's an exploration into the principles and methods that underpin the creation of gardens that can withstand adversity as well as contribute to the sustainability of the planet.

CREATING A RESILIENT GARDEN

There are no shortcuts when you are trying to create a resilient garden. It takes time to build a healthy soil and attract the diverse life that you need. I bought a small farm 20 years ago that had been managed conventionally, with regular applications of fertiliser and pesticides. Many of the soils were compacted from the use of heavy machinery. I put the land into organic conversion and started the slow process of improving the soils, as well as

Introduction

boosting biodiversity, creating new habitats, planting trees and letting areas rewild.

It's these principles that I have been applying to my walled garden, albeit on a much smaller scale, helping it to become more resilient and better able to cope with all the challenges that climate change will pose in the coming years, be it flood, drought or a myriad of pests new to our shores.

I studied botany and ecology at university and my research focused on the restoration of habitats, so I automatically look to nature for ideas. The one habitat that has influenced me more than any other is the rainforest – it never fails to impress me with its immense diversity and incredible stability. When you study ecology, you learn that the more stable habitats are also the most complex and that they have the most resilience to change. It is within these habitats that you see the most amazing relationships build up between plants and animals. We need to treat our gardens as a complete system and make use of ecological relationships that have evolved over millennia. Plants have evolved alongside their pollinators and pests, and they have developed mechanisms for dealing with these pests. They work with beneficial insects that attack the pests by supplying them with food in the form of pollen, nectar, root exudates, nutritionally rich leaves, seeds and fruits. You get even more complex relationships in the soil, where beneficial bacteria and fungi will live on the surface of roots and protect them from attack by pathogens.

Understanding your local environment is important, too. Learning about a plant's ecology – where it comes from, what conditions it needs, whether it's insect- or wind-pollinated – will help you choose the place in the garden where it is going to thrive. A Mediterranean plant, for example, which loves heat and free-draining soils, is never going to thrive on a heavy clay soil that gets waterlogged in winter.

I love sitting in my garden and just watching. Observing is just as important as the actual doing. I make notes and take photos so I can record what is working and what is not, see the changes from

Introduction

year to year, and in time, from decade to decade. So, I hope this knowledge together with my style of gardening, which is more hands off, more freestyle than neat, with plenty of biodiversity will lead to a healthy, resilient growing space.

SUSTAINABLE AND REGENERATIVE

The concept of resilient gardening is closely related to the principles of regenerative agriculture, permaculture and sustainable gardening. So, to start with, I am going to briefly describe these principles as I will come back to these principles throughout the book.

The aim of a sustainable garden is to minimise any negative impacts on the environment while promoting the health and well-being of the garden and its surrounding ecosystem. This is achieved through revitalising soil health, reducing waste, conserving resources such as water, avoiding peat, reducing the use of materials such as concrete, sand and gravel, managing pests and diseases in an environmentally friendly manner, and reducing the use of chemicals, such as fertilisers and pesticides.

Regenerative gardening is perhaps more proactive by looking to improve and restore ecosystems and to create systems that improve the environment, promote soil health and biodiversity. There is a focus on improving soil health through increasing biodiversity, enriching the soil with organic matter, making good use of water and helping to reverse climate change. A regenerative gardener might use practices, such as no-dig gardening, agroforestry, and permaculture, to enhance the overall resilience of their garden ecosystem.

There is considerable overlap with permaculture, a sustainable system of living created by observing natural systems. The term permaculture was first coined by the Australian Bill Mollison who, in his 1978 book *Permaculture One* written with David Holmgren, defined it as: 'The conscious design and maintenance of agriculturally productive systems which have the diversity, stability, and resilience of natural ecosystems. It is the harmonious integration of the landscape with people providing their food, energy, shelter and other material

Introduction

and non-material needs in a sustainable way.' In fact, permaculture encompasses regenerative, sustainable and resilient systems.

WHAT DOES A RESILIENT GARDEN LOOK LIKE?

I am often asked to describe a resilient garden. For me, it starts with plant diversity – lots of it! That means different species and varieties, as well as plant types in the form of annuals, perennials, bulbs, trees and shrubs. There is diversity, too, within the vegetables you grow and also temporal diversity, with crops at different stages of their life cycle. It's important that there is a patchwork or mosaic of habitats, such as long and short grass, herbaceous beds and shrubs, walls and fences, fruit trees, ponds and rills, and some wild corners, which all help the garden to support a diverse array of wildlife. In addition, a resilient garden is also about maintaining a balance in the soil, so that the plants and soil life can combat disease and keep pests at bay. The aim is not to get rid of them completely, but to keep them at a low level so there is little or no damage.

Gardening methods are important too. I look after my soils by not digging and by mulching and cover cropping to protect the surface of the soil. I like to operate a closed loop system as much as possible, so everything is composted and recycled back onto the beds. In the vegetable beds, I opt for more polyculture, i.e. the cultivation of multiple crops in the same space, rather than have neat and tidy rows of one type of vegetable. I grow lots of companion plants and I am pretty relaxed about the arrival of pests, as I know I have a good population of natural predators on hand. Like many vegetable growers, I save seed and, going forwards, it will become increasingly important to swap seed to boost the genetic diversity of our plants. We mustn't forget water resilience too. This is something I have been putting a lot of thought into as I design new areas of my garden, installing water-harvesting systems and thinking about where surface rainwater will drain, as well as choosing plants wisely.

Introduction

All in the soil

One of the keys to this diversity is a healthy soil – it underpins everything we do in the garden. When we improve our soils, for example, we sequester more carbon, which helps in the battle against climate change. Our soils will retain more water, improving the local water cycle and reducing erosion. Plus, a healthy soil will help us to create something that is vital to our own well-being – healthy food.

The idea that a healthy soil is important to the health of crops is not new – generations of gardeners around the world have made this link. One of the founders of the organic movement, Sir Albert Howard (1873–1947) wrote many books on the subject of soil health and composting based on his 25 years of research in India. Louise Albert, his second wife, wrote of his work in *Sir Albert Howard in India*:

> A fertile soil that is teeming with healthy life in the shape of abundant microflora and microfauna, will bear healthy plants, and these, when consumed by animals and man, will confer health on animals and man. But an infertile soil, that is, one lacking sufficient microbial, fungous and other life, will pass on some form of deficiency to the plants, and such plants, in turn, who pass on some form of deficiency to animal and man.

In the USA, author and publisher J.I. Rodale was inspired by the work of Albert Howard. Suffering from various ailments, he believed that his diet affected his own health so, in 1940, he bought a farm in Pennsylvania to start growing food without chemicals and coined the term organic. In 1947, he founded the Soil and Health Foundation, which later became the Rodale Institute. He understood the link between healthy soil and healthy food and ultimately, healthy people.

Research shows that fresh produce provides people with calories but not the essential minerals and vitamins needed for health. Nutrient density, the measure of the amount of nutrients you get in your food for the number of calories it contains, has declined greatly over the last 70 years. Since 1950, levels of

Introduction

calcium, iron and vitamins B and C in particular have fallen. For example, according to the USDA (US Department of Agriculture) broccoli in 1950 had calcium levels of 12.9 mg/g, but by 2009 this had fallen to 4.4 mg/g. This is a direct result of intensive farming where soil is tilled, fertilised and treated with pesticides, producing crops that have greater yields, but which lack nutrient content. It highlights how important it is for us to have healthy soils so we can grow truly healthy vegetables in our gardens and allotments.

YOUR OWN RESILIENT GREEN SPACE

Private gardens and allotments in Britain are estimated to cover an area of around 4 million hectares (10 million acres); that's more than all of the country's nature reserves combined. Of this space, around 500,000 hectares (1,240,000 million acres) are classed as urban residential gardens, so they are critical in providing habitats for plants and animals and creating green links between public open spaces and farmland, as well as for producing more home-grown fruits and vegetables.

If all of us did our bit and created a more diverse, healthy ecosystem in our own green spaces, be they large or small, or even community owned, we would be contributing to the overall goal of a much healthier environment.

In the chapters that follow, I will show you how you can work with nature to build soil health, grow a diverse selection of plants to create more habitats and attract more wildlife, manage resources, especially water, in a sustainable manner to create a garden or allotment that not only survives, but thrives in the face of adversity. It doesn't matter about the size of your plot, whether you have just taken over a new garden or allotment, are a seasoned allotment keeper or a novice gardener with dreams of a thriving plot, there are plenty of ideas in the following chapters to help you create a more resilient growing space which will be better placed to cope with the stresses that climate change is sure to throw at us in the years to come.

PART 1 Building a healthy soil

CHAPTER 1

Soil basics

Soil is the most vital asset of any garden or allotment, the unsung hero that serves as the linchpin for plant growth and overall ecosystem health. Its paramount importance lies in its multifaceted roles: as a reservoir of nutrients to provide essential elements crucial to healthy plant growth, as a medium for root support, and as a regulator of water retention and drainage, ensuring plants receive the right balance of moisture.

Beyond its physical properties, soil is a thriving ecosystem hosting an array of microorganisms that contribute to nutrient cycling, organic matter decomposition and even disease suppression, while soil's rich biodiversity enhances the overall health of the garden. Furthermore, soil plays a pivotal role in carbon sequestration, aiding climate-change mitigation by capturing and storing carbon from the atmosphere.

Soil is not a static medium, but a dynamic entity that, when understood and cared for, becomes the foundation upon which a thriving growing space is built, making it the cornerstone for cultivating resilient green spaces.

HOW HEALTHY ARE OUR SOILS?

Around the world, agricultural soils have suffered from the degradation caused by unsustainable practices, such as overuse of agrochemicals, poor irrigation, and overgrazing, which has led to

Building a healthy soil

nutrient imbalances and erosion. The result is the progressive deterioration of soil, which affects its ability to support plant life and ecosystem functions.

So, have our gardening practices been improving or harming our soils? It seems that many of the country's allotmenteers have actually been doing quite well. In 2014, scientists at Sheffield University found that allotment soils were significantly healthier than soils on intensively farmed land. Taking samples from allotments, parks and gardens in Sheffield and the surrounding farmland, Dr Jill Edmundson found allotment soil to be generally healthier and with less compaction than farmland soils. Of the plots sampled, the allotment holders had worked their plots on average for 5 years and a few for over 15 years – plenty of time for their practices to have an effect. Nearly all composted their green waste, and most allotmenteers added manure and fertilisers such as chicken manure, fish, blood and bone, tomato feed and liquid seaweed. The study showed that it was possible for small-scale urban food production to take place without the loss of soil quality associated with conventional intensive farming.

However, the same may not be true for all garden soils. Go into any garden centre and you will see shelves stacked with a huge variety of chemicals for use in the garden – lawn treatments, fertilisers, fungicides, pesticides, and more. They may provide the results craved by many gardeners but, in the long term, all they do is damage the soil and its inhabitants and create imbalances. Before you know it, more chemicals are needed to maintain that control and, from then on, it's a downward spiral. I talked to one gardener who had just moved house and then discovered her new soil to be in dire straits. Superficially, the garden appeared well looked after and healthy: the shrubs were pruned, the beds were full of interesting plants and the lawn was immaculate. But the perfect, weed-free lawn was a warning sign. Clearly it had been fed, mown and trimmed, but the removal of a sod of grass showed that the grass roots hardly extended into the thin soil beneath. Despite the outward impression, the garden soil was highly degraded, mirroring what we see in intensive agriculture, a consequence of relying heavily on synthetic fertilisers and pesticides rather than adding compost and mulch to build up the organic matter in the soil and feed the soil microbes.

The problem with fertilisers

When you add organic matter to healthy soil, microbes kick into action, breaking it down and releasing nutrients in a form that plants can use, whereas adding a synthetic fertiliser to your soil upsets the balance of soil life. When the fertiliser label says: 'readily available to plants', it means just that – it is in a form that plant roots can take up straightaway. The nutrients in the fertiliser bypass the soil food web and, instead, go directly to the plants' roots. I describe these instant nutrients as a 'plant-style ready meal'. The trouble is that continued use of these fertilisers results in many specialist microbes being put out of work and they then disappear from the soil community. The longer you continue to apply fertilisers, the lower the diversity of soil microbes becomes. Once the soil microbes disappear, the soil's ability to recycle nutrients decreases and the plants become dependent on fertiliser to supply all their nutritional requirements. The gardener must continue to feed the plants and the soil becomes nothing more than an anchoring material and water reservoir.

Of course, it's not just fertilisers that are causing problems. Pesticides, fungicides and herbicides can do just as much damage to soil life, disrupting the complex food web and putting things out of balance, and that is when pests and pathogens can start to get the upper hand. Fungicides, for example, are designed to kill pathogenic fungi, but most fungicides are not specific to a particular fungus. Consequently, an application of fungicide to treat one specific fungal infection may lead to the chemical draining into the soil and affecting many other soil fungi.

WHAT'S IN YOUR SOIL

So, before we think about ways of regenerating soil, here's a quick overview of soil and its components. Basically, it's a mix of minerals, organic matter, air and water, and living organisms.

Minerals are the sand, silt and clay particles that make up about half the soil. They're the breakdown products of the bedrock that lies below the soil and range in size from the smallest clays to the coarsest sand particles. The proportion of sand, silt and clay in a soil determines its texture, water-holding ability, drainage and fertility. Loamy soils are highly desirable because they have a balance of all three.

Organic matter refers to all those materials and matter, both plant and animal, that were once living. It includes organic inputs, such as composts and manures, all at various stages of decay. Such matter is the main food resource for all soil life, being about half carbon, together with nitrogen, phosphorus, potassium, magnesium, calcium and micronutrients, which are released as the organic matter decomposes. Some of the organic matter may be a recent addition, while some may have been present in the soil for decades or even hundreds of years.

Air and water content is variable. For well-aerated soil, half the volume is made up of air spaces (or pores). However, the proportion occupied by air and water varies considerably – when it rains, for example, water drains through the soil, pushing air out of the pores. Water contains dissolved nutrients which are needed by plants.

Soil life comprises a highly diverse mix of bacteria, fungi, protozoa, nematodes, springtails, mites and larger animals, including earthworms and beetles. Such is the diversity of life

Soil basics

that many soil species are still unknown and even those that are known are little studied (see 'The rhizosphere', page 16.)

Soil organic matter

Soil organic matter (SOM) is a key indicator of soil health. It improves fertility and helps soil particles clump together to form aggregates, which improves the soil's water-holding capacity, permeability and drainage, and reduces the risk of soil erosion. The presence of organic matter not only stimulates soil decomposers, such as bacteria and fungi, but the whole soil food web also responds to organic matter.

The amount of SOM in your soil depends on how much organic matter you add and the rate at which it is broken down. It's also affected by soil texture and the local climate. Organic matter tends to be concentrated in the topsoil and decreases with depth, unless you dig and turn your soil. However, not all sources of organic material are equally good at boosting the organic matter in soil. The best improvements come from the addition of bulky farmyard manure, which boosts soil biology and structure, with the next best being compost.

What is a good amount? Organic matter in a healthy soil will be at least 3 per cent, ideally more, although it depends on soil type and rainfall. On light, sandy, arable soils in low rainfall areas, such as East Anglia, farmers aim for at least 2–3 per cent organic matter, whereas in the higher rainfall areas of South West England the target is 4–6 per cent. Farmers with a heavy clay soil aim for 4–9 per cent, depending on the rainfall.

It's crucial therefore that you continue to add organic materials to the soil on your plot, ideally as a layer of mulch. Some organic matter does not break down, so it becomes a permanent part of the soil, holding onto nutrients and preventing them from leaching away. And, of course, since organic matter also acts as a carbon store, by improving your soil you are sequestering more carbon and helping in the battle against climate change.

Building a healthy soil

GOOD SOIL STRUCTURE

Soil scientists often talk about 'good soil structure', which describes the make-up of soil particles and how they clump together. We want a soil with an open, crumbly structure and plenty of pores to facilitate the movement of air and water, as well as vertical cracks that allow roots to grow down deep. And it's biological activity that's behind a good crumb structure. Soil microbes secrete a glue-like substance called glomalin that sticks the soil particles together to form crumbs. If you look at soil with

The theory of soil

New research by the Rothamsted Research group, which they call 'the theory of soil', has thrown more light on soil structure. They found that in a healthy soil the relatively low levels of nitrogen encourage microbes to produce glomalins and other glues, which boost soil aggregation and help to create a more resilient soil that is better able to cope with drought and flooding. The research showed that the shift from the use of carbon-rich manures to artificial fertilisers high in nitrogen and phosphorus changed the way microbes used nutrients. They produced less glue, which meant that soil structures changed, with fewer pores and less oxygen. Now, decades later, intensively farmed soils are very different to a grassland soil and are more reliant on fertilisers.

a good crumb structure, you can see that the crumbs are separate and there are lots of roots covered in soil.

When it rains, water drains through the soil to deeper levels, filling the air spaces. Later, when the weather is dry, the water evaporates, although some remains as a film around the particles to be taken up by plant roots. A soil rich in organic matter has more pores and this leads to more water storage and better resilience in times of drought. In contrast, a soil that has suffered from compaction will break into large angular pieces and, although there may be roots, there won't be many, and they are usually white rather than covered in soil.

Have you ever dug a hole to plant a shrub and noticed that your soil looks dry, even though there have been weeks of rain? Maybe it's moist at the top, but dry underneath. When you see this, you'll know that you have a problem because the rain has been unable to percolate down through the soil. Instead, it's either pooled on the surface or run off onto a path or into a ditch or watercourse. This means that the soil is more vulnerable to drought because the water has not moved deep into the ground to replenish the long-term stores. A compacted soil has fewer pores and fissures, which makes it more difficult for water to penetrate and for soil organisms to move around.

SOIL pH

Soil pH is a measure of the acidity or alkalinity of soil, on a scale of 0 to 14, with a pH of 7 being neutral. It can be measured using a pH meter or indicator papers. The pH of soil affects both the availability of minerals and soil life. Loamy soils, especially those used for growing crops, tend to be in the neutral pH range, which means that all the soil minerals are available to the plants. But, if the pH increases or decreases, their availability is reduced. Also, the pH of a soil may alter over time – for example, if crops are continually removed and the nutrients not replaced. The continued use of artificial fertilisers will also alter soil pH, as will the

Building a healthy soil

regular application of naturally acidic composts and manures, such as a poultry manure. Soil life is affected by pH, too. Acidic soils tend to support fewer fungi and bacteria, and important groups of animals, such as earthworms and nematodes, disappear. Plants are also affected by soil pH. Plants such as azaleas and rhododendrons prefer an acid soil and cannot tolerate chalky soils. In contrast, there are many plants that thrive on chalky soils, so it is important to make sure that the soil pH suits the plants you grow.

SOIL TESTS

It's easy to do a few tests to assess the health of your soil and identify any issues that you need to address, such as soil organic matter or drainage. Use the results as a baseline and repeat the tests in a year to see if your soil is improving.

Visual Evaluation of Soil Structure (VESS) To carry out this evaluation, you need to download a VESS sheet from the internet and go through the series of assessments. Dig out a block of soil the width and depth of your spade, levering it out of the ground and placing it on a shallow white tray. Follow the assessments on the sheet: look at how easily the soil breaks up in your hand, the shape of the block, the colour and smell, etc. Check out the roots, too. Brown roots indicate lots of microbial activity since the soil particles stick to the roots because of all the glomalin. It's the start of good crumb structure. If the roots are white, however, there are fewer microbes at work and so you will need to add more organic matter.

Water infiltration test This will show how easily water can percolate through the soil and whether you need to do some remedial work. Dig a square hole, about 30cm (12in) across and 30cm (12in) deep. Fill the hole with water and allow it to

Soil basics

drain away overnight to make sure the soil is saturated. The following morning refill the hole with water and record how long it takes for the water to drain away. Ideally, you are looking for a soil that allows water to drain away in 2–4 hours; anything outside this range means you need to take action. If times are extended, perhaps by as much as 10 hours, you probably have compaction which is stopping the water from draining away. A slow-draining soil is liable to become waterlogged and this can kill plant roots. If the water drains away in an hour or less, the soil is very sandy and in dry weather the plants will need a lot of watering. Organic matter is usually the answer to better drainage in both circumstances. Adding organic matter to a sandy soil will improve its structure and allow more water to be retained. Similarly, mixing organic matter into a heavy clay soil will also improve its structure, helping water to move through the soil more quickly and reducing drainage times. You can add as much as a third (by volume) of organic matter, either placing it on the surface or turning it in. If you have particularly heavy clay, the soil will benefit from the addition of grit as well.

Slake test This is used to test for crumb structure – it looks at how quickly the soil breaks up when placed in water: the slower the soil breaks up, the more organic matter is present. Bring a sample of soil inside and carefully remove three pea-sized lumps of soil. Place each on a teaspoon and leave overnight to dry out. Fill a shallow dish with rainwater and, using the teaspoons, gently place the lumps in the water, spaced well apart. Watch what happens over the next minute. A soil with good structure holds together even after a full minute, whereas soil with a poor structure will collapse almost immediately. This is because soils with a better crumb structure have more organic matter, which binds the particles, helping the soil to hold together during wet weather. They can maintain their structure, providing air and water for plants.

Building a healthy soil

Assessing earthworm populations Earthworms are indicators of soil health. Their numbers decline if the soil becomes too acidic, is turned or rotovated, or becomes compacted or waterlogged. But they increase as the soil organic matter increases. The ideal time to count earthworms is during a warm, wet spell in late spring or early autumn when they will be present in the upper layer of soil. Dig a pit measuring 20 × 20 × 10cm (8 × 8 × 4in) and spread the soil on a shallow white tray. Break up the soil and collect the earthworms. The simplest assessment is a count of the adults (the ones with a saddle) and juveniles. Once complete, put everything back in the hole. A good count is 10 to 15 adult earthworms. But you can go one stage further and identify the different species using online guides.

THE RHIZOSPHERE

There is a whole food web in the soil around plant roots called the rhizosphere. This is an ecosystem in its own right, where the plant roots and organic matter create an environment to support and shelter soil life. There's much to learn about the rhizosphere. We haven't identified many of the organisms that live in soil, let alone understood their role and how they interact. Among the newly discovered soil organisms are 'archaea', life forms that were originally discovered in the hot geysers of Yellowstone National Park in the US and initially thought to be bacteria. These unique life forms, along with bacteria and fungi, are food sources for predators such as nematodes and protozoans which, in turn, provide food for larger predators.

Plants have a critical role to play in soil health. You will read in gardening books how plants take up nutrients from the soil via their roots, but there is little mention of the fact that there is a two-way flow, as plants exude as much as 40 per cent of their photosynthetic production from their roots into the soil to feed the microbial community around the roots. These exudates vary with the health status of the plant. A healthy plant produces a more diverse selection of substances, including complex carbohydrates and proteins, and this leads to more soil microbes and a healthier soil. In contrast, a less healthy plant tends to release more simple sugars which encourage pathogenic organisms. So, the relationship between the soil and plants works both ways.

HARDWORKING INHABITANTS

Soil inhabitants help to improve soil structure and take all the waste materials from plants and animals, recycling them via the various nutrient cycles. Through their actions, the amount of humus (the end product of decomposition) in the soil increases. This is a stable material that acts as a nutrient reservoir for plants, helping grow nutrient-dense crops that are so critical for our health. In addition, some microorganisms have the ability to digest clay, silt, sand and even shells, to further boost soil mineral content.

The fungal world

Fungi, alongside bacteria, are key soil inhabitants, ranging from simple microscopic yeasts to highly complex structures such as toadstools. Many fungi are involved with decomposition and nutrient recycling. A few are plant pathogens and there are some that have a mutually beneficial relationship with plants. Most fungi are made up of long threads called hyphae that extend through the soil. You see these white threads growing through leaf litter, compost or piles of woodchip. Collectively, hyphae form a network called a mycelium. Most fungi feed by pouring enzymes onto organic matter, breaking up the large complex compounds into smaller soluble ones that can be absorbed by their hyphae.

Mycorrhizal fungi have mutualistic (symbiotic) relationships with plants: the fungi provide plants with nutrients and water, while the plant provides the fungi with sugars. A mycorrhizal fungus has part of its structure within the plant root and part in the surrounding soil, greatly increasing the area from which the plant can obtain nutrition. As scientists have learned more about soil fungi, they have found that the fungal-plant relationship is not one to one but multifaceted, with fungi having mutualistic relationships with many different plants over a large area. Not only do nutrients pass from plant to fungi, but they also pass from plant to plant via the fungal network. In fact, scientists have found that nutrients may pass from a dying plant to a living one, and from a mature one to a young one, via the hyphae. It was leading mycologist David Read who referred to this immense and truly fascinating network in the soil as the 'wood wide web'. This web isn't just a network for distributing nutrients. It's also a communication network that alerts a community of plants to the presence of pests or pathogens. It can even protect plants from underground herbivores that would eat their roots, help with the distribution of plant nutrients to soil organisms and enable a systemic response by the entire ecosystem. We now also know that there are mutual relationships between fungi and soil bacteria. Indeed, many bacteria use fungal hyphae as paths to move around the soil. But it goes even further. Martin Pion, when working with other researchers back in 2013, found that one species of morel (a type of mushroom) actually 'farms' bacteria, rearing them on its exudates and then harvesting the bacterial sugars.

There are two groups of mycorrhizal fungi and these differ in the way they interface with plants. Ectomycorrhizal mycorrhiza, which are associated with many trees, surround the root like a sheath, while the more common endomycorrhizas extend their hyphae into the root. But, despite these structural differences, their role is the same. Around 90 per cent of plants have associations with mycorrhizal fungi. However, there are two notable exceptions: the Brassicaceae (cabbage) and Amaranthaceae (formerly Chenopodiaceae) (spinach, quinoa, orache) families. The lack of a mycorrhiza in these important crop plants has implications for

Soil basics

their cultivation which I discuss later (see 'Brassicas – a special case?', page 98).

When we think about improving the health of our soils, central to our task is making sure we have a healthy fungal network. One way we can do this is not to disturb our soils as digging (tillage) rips the network apart.

Adaptive nematodes

Soil nematodes are microscopic, thread-like organisms that swim in the film of water surrounding soil particles. They are highly adapted and assist with nutrient cycling by feeding on organic matter, bacteria and fungi. A small number cause plant disease, while some are parasites of other nematodes, snails and slugs, and insects. Some nematodes can protect plants from pathogens, including those nematodes that attack plant roots, so it's not surprising they are useful biocontrol agents (see 'Nematodes', page 119). Soil scientists look at the range of nematodes in a soil sample and the balance of beneficial and pathogenic species as an indicator of soil health.

Earthworm engineers

Earthworms are ecosystem engineers, altering soil as they burrow and moving particles around. They improve soil structure, which leads to better aeration, drainage and water-holding capacity. They eat fallen leaves and organic matter, along with any bacteria and fungi living on these, helping the process of recycling and producing faeces that improve the soil. There are four main groups:

Anecic are deep-burrowing earthworms, more than 8cm (3in) long and dark red-brown in colour. They make vertical burrows and leave worm casts on the surface.

Endogeic are small-to-medium-length earthworms, pale in colour, and they tend to curl up when handled. They create a

Building a healthy soil

network of horizontal burrows in the top layer of soil, improving soil crumb structure and drainage.

Epigeic are small, dark red earthworms found in leaf litter.

Compost earthworms are medium-sized, bright red and stripy, typically found in compost heaps and areas with a lot of decaying vegetation on which they feed. They are rarely found in soil.

CHAPTER 2

Regenerating your soil

There's lots of things you can do to improve the resilience of your soil, and the easiest and most effective one is to feed it by adding some organic matter, such as compost or farmyard manure. Not only do the microbes love this, but organic matter also improves the soil's water-holding properties, providing better aeration and drainage. Organic matter also helps soil structure by allowing plant roots to spread more easily through the soil. You can also improve the resilience of your soil by:

- Minimising soil disturbance – so as not to disturb the soil food web.
- Maximising crop diversity – since different crops have different rooting depths and support a wider range of biodiversity, both above and below ground.
- Providing year-round soil cover – to protect soils from wind and water erosion, and also reduce water loss.
- Keeping a living root in the soil for as long as possible – plants produce root exudates, substances such as sugars, amino acids and organic acids released by roots, which benefit microbes and support soil health.

- Integrating livestock – this promotes species diversity, from microbes to mammals, and puts organic matter in the form of manure back into the system. In a garden, this can be achieved by keeping poultry and adding farmyard manure.

NO-DIG GARDENING

Look through any gardening book written 30 or 40 years ago and you will find plenty about digging – using single and double spade depths, with diagrams showing you how to turn the soil over. My grandfather liked digging, and I think he gained a lot from the therapeutic benefit, the satisfaction of stepping back after all that hard work and admiring the freshly turned soil without a weed in sight. He may have felt that he had achieved something, but I don't think his soils were any healthier or more resilient for all that effort. When I first started writing about soil, more than 20 years ago, there was hardly a mention of 'no dig', or 'no till' as it's called in North America. This is when soil is not dug or turned. Instead, compost is simply spread over the top of the soil. It's a method made popular by Charles Dowding that's gaining traction now, especially among the 'grow your own' community, and I'm even beginning to hear about diehard spade-wielding gardeners taking this approach, too.

No dig is a style of soil management that mimics nature. In a forest ecosystem, leaves, twigs and fruits drop to the ground and cover the soil with a protective layer of litter which is broken down by the soil life. This replenishes the soil with nutrients and also protects it from adverse weather. I have used no dig myself for many years, first by abandoning the spade and simply loosening the soil with a fork, and then moving on to full no-dig methods (see 'Simple no-dig guidance' box). I don't claim to be 100 per cent no dig, though, as there are times when I use a fork, but most of my growing space is managed on no-dig principles.

Regenerating your soil

When you turn soil, you expose it to air and the carbon locked up in the soil reacts with oxygen to create carbon dioxide. So, digging actually adds to global warming. It also brings weed seeds to the surface where they germinate, which creates more weeding. No dig, on the other hand, really benefits soil life. Just think of earthworms – being chopped in half by a spade is never going to help the soil. But the real benefit of no dig is to the soil fungi because the delicate 'wood wide web' of hyphae extending through the soil is easily broken. I walk on my soil, too, the cardinal sin of gardening, but having not been turned, my soils are not so easily compacted and so are far more resilient. Plus, no-dig methods save so much time! Time that can be better spent in the garden building fertility and growing plants.

Simple no-dig guidance

- Every autumn or winter add a layer of organic matter (compost or mulch) to the soil. You can sow a green manure (see 'Cover crops', page 32) on cleared vegetable beds and mulch around the perennials in herbaceous borders.
- Don't pull out crops at the end of the season, but just cut them off at soil level. The roots remaining in the soil will add to the organic matter and help to hold the soil together during the winter months.
- Hoe the surface to remove weeds.
- Disturb the soil as little as possible when planting or pulling up deep-rooted weeds, potatoes and root crops.

Building a healthy soil

COMPOSTING

I am always surprised by the number of bags of green waste put out for collection or taken to recycling centres, even by gardeners. Why would gardeners want to throw away such valuable materials? If you are working to regenerate your soil, this waste should really be composted in the space in which it was produced and then returned to the soil.

During the composting process, bacteria, fungi and other microbes break down complex molecules into simple compounds, such as nitrates, that plants can use. It's very similar to the process of digestion in our own gut. The end product of all this decomposition is humus – a stable, brown, crumbly material that is rich in nutrients.

Slow, semi-static composting

There are lots of ways to compost garden waste, but I suspect most gardeners use the slow, semi-static method of simply tipping garden and kitchen waste onto the compost heap. Occasionally, they give it a turn, but they are pretty hands off and let nature take its course. This method works but takes a year or more, and you often end up with a compost full of weed seeds. Furthermore, the compost may not reach temperatures high enough to kill pathogens (see 'Faster composting'). I confess that I use this method myself, as I am short of time and don't have huge amounts of material to add at once, but I do make sure that I add both brown and green materials (see 'Table 1. Composting materials'), ideally around 60 per cent brown to 40 per cent green. I'm usually short of brown materials in summer, so I keep a pile of woodchip and spare cardboard near the compost bins so that I can adjust the balance. More recently, I have taken to inserting vertical plastic tubes punctured with holes into the middle of the larger heaps, which helps to aerate the compost and make up for the lack of turning. Once the bin is full, I take the compost out and give it a mix, remembering to watch out for frogs, toads, slow worms or snakes that are often hiding among the compost.

Regenerating your soil

Shredding the material helps, too, since the shredded pieces present a larger surface area to the microbes and will therefore compost more quickly. It's well worthwhile saving woody prunings and having a shredding session. The resulting fine clippings

Table 1. Composting materials

	BROWN	**GREEN**
Carbon: Nitrogen ratio	Carbon-rich, i.e. high carbon to nitrogen (C:N) ratio, woody materials with lots of lignin that take time to compost.	Nitrogen-rich, i.e. low carbon to nitrogen (C:N) ratio, non-woody materials, often rich in sugars, that are quickly broken down.
Decomposition	Broken down mostly by fungi and actinomycetes that have the right enzymes to tackle lignin.	Broken down by bacteria but be careful not to add too many grass clippings as these can decompose into a smelly, slimy mush.
Examples	Bracken, shredded cardboard, paper and woody garden waste, fallen leaves, straw, wood ash, wood shavings, woodchip and sawdust	Coffee grounds, tea bags, comfrey leaves, green garden waste, grass clippings, hay, seaweed, vegetable peelings, citrus fruit and banana skins, pondweed, spent hops, wool fleece and urine

Note: Don't add the following to compost: dog faeces and cat litter, glossy paper, meat and fish (fresh or cooked), eggshells, grass clippings from lawns treated with weedkiller.

can be mixed with grass clippings, rotting apples or manure and added to the compost heap.

Faster composting

This is based on generating higher temperatures and these are the methods used by commercial composting companies that collect household green waste. You can compost more quickly if you generate more heat, with the added benefit that the compost is less likely to be full of seeds or pathogens. Many gardeners use a 'hot-bin', which allows them to compost using high temperatures in enclosed bins with little risk of attracting vermin to the garden; something that I know many urban gardeners are concerned about.

The method described here is based on one that I saw demonstrated by soil experts Urs Hildebrandt and Angelika Lübke-Hildebrandt who are both from Austria. The turnaround time is three months, if carried out in spring and summer. The aim is to produce a rich, crumbly compost that is full of nutrients, doesn't smell and has an abundance of life. I have tried faster composting, and it works! But it takes a lot of effort and you will need to have a fair amount of compostable material before you start to build the windrow (a long line of heaped material). The advantage, however, is that you can generate a large volume of compost quickly. You will need some compost fleece to keep the windrow covered and a compost thermometer with a long probe which you stick into the middle of the heap to check the temperature.

The principles described here can be adapted for use in a large garden and smallholding, or perhaps an allotment where a group of allotmenteers can work together. Aim to collect enough compostable materials to make a windrow measuring 5m long by 2m wide and 1m high (16 × 6 × 3ft) using the following materials:

- The mix of compost materials needs to be 60 per cent brown to 40 per cent green. The bulk of the

Regenerating your soil

brown can come from straw, woodchip, shredded woody prunings and cardboard, etc. Green materials include hay, fresh weeds, grass clippings and kitchen food waste, and farmyard manure, which will supply the microbes with food. But make sure you don't add more than 30 per cent grass or manure and, ideally, incorporate these 'hot' materials over a couple of days, otherwise the temperature can increase too quickly. Bales of spoilt hay and straw are perfect for providing bulk. Ideally, the bales should be soaked before adding to the windrow.

- If possible, add about 10 per cent by volume of clay or a loamy-clay soil, to help establish a clay-humus complex.
- 10 per cent by volume of finished compost (if available) to provide some microorganisms, so decomposition can start immediately.
- Some people like to add rock dust or biochar (see 'Biochar' box, page 29) at a rate of 2kg per cubic metre (4½lb per 35 cubic feet), which have been found to speed up the composting process, but this is optional.

Position the windrow on an area of compacted soil or solid ground and, ideally, on a slight slope so that water can drain away. Build up the windrow using layers of the different materials, starting with straw and hay, alternating the layers of brown and green, and ending with clay and compost. Adding the material in layers helps you to estimate the proportions. The layers then need to be well mixed, fluffing up the materials to get lots of air incorporated. Finally, water the windrow well to achieve a 55–60 per cent moisture level. As you reach this level, water will begin to run out of the bottom of the heap, and you will need to maintain this level of moisture during the composting process. When the windrow is complete, cover it with a compost fleece to trap the heat. The fleece lets air through, so it is better than plastic sheeting, but plastic will do if you don't have a fleece.

Check the temperature of the windrow daily. The temperature must not exceed 65°C (149°F), so turn the windrow when the temperature reaches 60°C (140°F). The aim is to achieve 58°C (136°F) for 10 days to ensure that all the pathogens are killed, so you may need to turn the pile as many as three times. If you find that the temperature doesn't increase enough, the mix may be too dry or there could be a lack of energy-rich food for the microbes. If this is the case, add some grass clippings or manure to give the mix a boost.

Eventually, as the energy-rich compounds that the microbes are feeding on are exhausted, the compost heap will stabilise and the temperature will drop. If done correctly, this composting process will ensure all the pathogens and weed seeds are eliminated and the nutrients will be locked up in the clay-humus complex (meaning they will not leach out when spread on the soil).

Bacteria- versus fungi-rich compost

By tweaking the materials used, you can produce a compost that is richer in bacteria or in fungi, which can be put to different uses. A bacteria-rich compost is one that has been made from a higher percentage of green materials, such as manures, grass clippings or hay. It is particularly good for mulching soils to grow vegetables (especially brassicas which don't have mycorrhizal fungi), flowers and herbs. A compost that is rich in woody material and has a high fungal component is good for mulching fruit trees and bushes, new hedgerows and perennial plants.

Hay and straw are useful composting materials. Hay is made from grass, cut when the leaves are full of sugars and proteins and then allowed to dry. It's a green material and, if used in compost, it will boost bacteria. Straw is made from the leaves and stems of plants that are in the dying phase of their life, so they contain far fewer sugars and proteins and more carbon. It's a brown composting material that will encourage fungi.

Regenerating your soil

Biochar

Biochar is a soil amendment. It's made in the same way as charcoal, by burning plant material at high temperatures without oxygen to create a black substance that is rich in carbon. The granules look solid but have a honeycomb structure, so they are porous and rigid. Mixed into soil, biochar provides many benefits: better aeration, less compaction and retention of soluble mineral nutrients, so there is less leaching. In addition, soil microbes shelter in the holes where they can hide from predators. It's estimated that the number of beneficial fungi and bacteria can be 10 times higher in soil enriched with biochar than soil without. Biochar holds onto water, too, so the soil's water-holding capacity is much greater. Research shows that the use of biochar around the roots of transplanted saplings leads to greater survival rates as it improves their ability to withstand drought.

I like to use a potting compost with added biochar to boost root growth. I simply mix in about 100g/3½oz to every 4 litres/1gal of compost. You can also sprinkle biochar over the top of the soil. I add some to my compost bins, too. This helps to 'biocharge' the compost, so it's full of beneficial microorganisms before it goes on the soil.

Building a healthy soil

Using poultry

Livestock dung enriches soils and helps to improve resilience, which is why many arable farmers are introducing sheep and cattle into their rotations. While I'm not suggesting you keep a flock of sheep in your garden, you may have the space for a trio of hens or ducks that will act as pest controllers and provide you with manure and, of course, eggs. My own poultry are housed on straw, and I add this nutrient-rich waste to my compost heaps. But I think one of the best compost heaps is that turned by poultry. If you have enough space in your poultry pen, toss in all your weeds and shredded garden waste. The chickens will have a great time pecking at seeds and insects, giving it all a good mix. Not only does this reduce the amount of feed they need, but they will be activating your compost heap with their droppings.

Fermenting compost

There has been a huge surge in interest in fermented foods recently, but it is also possible to ferment compost. This means breaking down the compost anaerobically (without oxygen) through a process called 'bokashi'. For this, you need an inoculant of Effective Microorganisms (EMs), such as lactic acid bacteria, yeast and photosynthetic bacteria. Gardeners have long been able to buy 'bokashi buckets' and bran inoculated with EMs to compost their food waste. I was a bit sceptical at first, as my only experience of anaerobically digested plant waste was a smelly heap of grass clippings, but I wanted to do something with my food waste that I couldn't put on the garden compost heap. The bokashi bucket has a tightly fitting lid and there's a tap at the bottom. You add your food waste to the bucket, mix it with inoculated bran, press it down into the bucket to get rid of the oxygen, and sprinkle with another layer of bran. Then the lid is closed tightly and the bucket left for up to 12 days. You can keep adding new layers if the bucket is not full up. The tap at the bottom is used to drain out any liquid, or 'leachate', which can go in with the compost. This process is really like pickling. At first, I

Regenerating your soil

Improving soil in a new-build garden

Many people with a recently built home will have garden soil that has suffered from the construction process – the 'garden' will have been trampled and compacted during the build and finished off with a thin covering of 'topsoil' over what is often just building waste. The new homeowner is left with pretty poor soil that usually suffers from poor drainage and pooling of water on the surface and is frequently devoid of any life. But this is now 'virgin' growing space. So where do you start on the regenerative process? What about digging? Normally, I would say don't dig. But when you are faced with compacted soil, you really need to break up the compacted material that is preventing water from draining. The soil needs to be forked over, which will also allow you to find the rubble that is sure to be there, too. If the amount of rubble is modest, you could use the technique used by many growers of inserting a broad fork and rocking it back and forth to create passageways through the soil, rather than actually lifting the soil, which can be backbreaking. However, if there is loads of rubble, you will have to dig the soil and then mulch – you're likely to need quite a lot of material, so I'd choose a woodchip mulch. This will help to improve a compacted soil, add organic matter, retain

> moisture and can often be obtained for free. You can also use well-rotted farmyard manure or a material like digestate. But remember, you are adding soil conditioners, so you don't need to buy bags of quality peat-free compost – save that for your propagation and containers.
>
> If you have a nutrient-poor soil, it will be perfect for establishing a wildflower meadow, as meadow flowers thrive on low fertility. Too much fertility and the more aggressive grasses will dominate the wildflowers.

thought the compost would be ready for use after 12 days, but this is not the case. It is just that after this period it can be added to the compost bin or spread over your garden beds to continue the composting process. Any bones must stay in the bin for much longer. You can't plant anything straight into bokashi compost, as it is very acidic, but if you leave it on the soil, it will continue to break down and you can plant into it after six to eight weeks. If you have enough food waste and don't want to put it in a local authority green waste collection bin or on your own compost heap, this is a good option.

COVER CROPS

Compost is great for boosting soil fertility, but many regenerative and organic farmers use cover crops or green manures to improve soil structure and fertility. These are fast-growing crops that are sown after the main crop has been harvested. They protect the soil over winter, reducing the risk of soil erosion and nutrient leaching,

Regenerating your soil

Table 2. Cover crop options

TYPE	EXAMPLES	FEATURES	WHEN TO SOW
Annual ryegrass (*Lolium multiflorum*)	Westerwold ryegrass	Fast-to-establish, deep-rooting grass that protects the soil and improves soil structure. Make sure you kill it off before it sets seed in spring by mowing it several times until it eventually dies. You can also cover with black plastic sheeting in late winter and leave in place until you are ready to plant the next crop.	Late summer and autumn
Brassicas	Mustard, fodder radish, fodder turnips	Fast-growing, deep-rooting plants to lift nutrients. They produce lots of biomass and help to inhibit soil pests such as wireworms and nematodes (see 'Biofumigation', page 141). They are killed off by frost. Be aware that brassica cover crops can harbour club root.	Late spring to late summer

Table 2. Cover crop options (*continued*)

TYPE	EXAMPLES	FEATURES	WHEN TO SOW
Buckwheat (*Fagopyrum esculentum*)		Fast-growing annual that lifts phosphate. The flowers attract a lot of pollinators. Don't let it seed, as this results in lots of seedlings the following year. Killed by frost.	Mid-spring to late summer
Legumes	White clover, crimson clover, Persian clover, fenugreek, lucerne [alfalfa] (*Medicago sativa*) and vetches	Nitrogen-fixing plants that boost soil nitrogen when they rot down. To build fertility, legumes can be left for 12 to 18 months before being terminated.	From spring to late summer
Phacelia (*Phacelia tanacetifolia*)		Easy-to-grow, deep-rooting, nitrogen-lifting cover crop that also makes a good pollinator crop. Will self-seed everywhere if allowed to flower.	Sow in autumn and leave over winter, or sow mid-season to fill in gaps.

adding bulk, improving water-holding and drainage, and helping to increase biodiversity above and below the soil. When the farmer is ready to sow the next crop, they are killed off and left as a protective mulch, which can boost soil organic matter and disrupt the pest life cycle.

But how cover crops are terminated is causing problems for the regenerative farmer. Farmers often kill off the cover crop by turning it into the soil with a plough, but tilling is not an option if you are rebuilding your soil. Another method is to kill the cover crop with glyphosate weedkiller, an option that is not available to the organic farmer, so farmers are instead developing methods that use specialist mechanical crimping machines to break the stems, killing the plants and leaving a mulch on the surface.

Cover crops have the same benefits and challenges for the gardener. Sowing cover crops is easy but, just as for the regenerative farmer, it's how you kill them off that can be tricky for the no-dig gardener. One way is to use frost-sensitive species that grow through late summer and autumn but are killed by winter frost. The plants die and create a mulch over the surface of the soil. Another option is to mow the crop close to the ground and then cover the area with cardboard or plastic to smother and kill the plants. You can even plant directly into the resulting mulch, although some say that there will be too much of a slug risk. However, the mulch is also home to spiders, beetles and other predators that use it for cover over winter.

MULCHING

Another way to keep soil covered is to use a mulch. Mulch is a layer of organic matter, compost or similar material, which is spread over the surface of the soil to act as a barrier against water loss and heat, reduce weeds, improve soil structure, provide a habitat for animals, and more. It really is essential to soil health. Most gardeners are diligent about mulching their newly planted trees and shrubs to give them a good start, but perennial beds and

Building a healthy soil

vegetable plots tend to get overlooked. Mulches can be applied all year, ideally after rain so that they can trap the moisture in the soil. A good time to mulch is autumn, when crops have been harvested and you don't want to leave the soil exposed over winter. A thick layer isn't required, just a few centimetres will do, but don't mulch right up to the trunks and stems of plants as this can lead to rot.

A mulch can also make it difficult for pests. Research has shown that covering bare soil with a mulch makes it difficult for flying insects, such as aphids, flea beetles and leaf hoppers, to distinguish the crop from the mulch, so it can reduce pest damage in spring while at the same time encouraging natural predators. There are lots of materials that you can use as a mulch:

Compost A thin layer of your own garden compost is ideal, but don't add more than a couple of centimetres in depth.

Grass clippings These need to be spread thinly, otherwise they rot down and create a foul-smelling, rotting grassy mat. Usually, I spread fresh clippings out to dry first and then apply them to my beds.

Leaf mould This is one of my favourite materials. Collect fallen leaves and let them rot down into leaf mould (see 'The value of leaf litter', page 38).

Straw and hay A thick layer of straw or hay is great for trapping moisture and supplying nutrients, but it can encourage slugs, snails and spider mites.

Mineralised wheat straw and maize biodigestate These are commercial products that you can buy to mulch your flower and vegetable beds. These materials reduce weed growth and enrich the soil, and they last for several years. They may even deter slugs and snails and, when used regularly for a number of years, gardeners have reported improved plant health and soil structure.

Regenerating your soil

Wool As a smallholder, I have a small flock of sheep and once a year, when they are sheared, I get a new supply of wool, and it's easy to put this down to create a barrier. I think the wool (especially the really smelly daggings) deters rabbits, possibly deters slugs and snails when dry, and is also a source of nutrients. I like to use wool as a mulch around container plants and newly planted trees and shrubs, to create a thick barrier to retain moisture, and around transplanted brassicas to deter cabbage root fly. The downside is that white wool is not particularly attractive to look at.

Mulching materials don't have to be organic. Gravel is a versatile and practical mulch for the garden and delivers a range of benefits, especially when it comes to reducing water loss from the soil. A thick layer of gravel over the soil minimises evaporation and retains the water for the plant roots. The gravel acts to suppress weeds too, while in summer protects the soil from the heat of the sun. A sustainable alternative to gravel is crushed recycled aggregate from the construction industry. I've used a thick layer of this material, obtained when we demolished a farm building, to create my climate-resilient gravel garden.

A living mulch

I am particularly fond of using a 'living mulch'. As the name suggests, this is a covering of plants, sown either under the main crop or intercropped with it. I grow white clover under crops such as courgettes and brassicas. The white clover (*Trifolium repens*) is sown first and once this is established I plant out the main crop. The clover fixes nitrogen, helps to smother weeds and keeps the soil protected. Its growth rate decreases as the main crop gets larger and there is more shading but, once the main crop is harvested, the clover grows on through the winter, helping to improve the soil fertility. Its flowers attract both pollinators and predators. I let the clover grow on through another year to help build up fertility. I terminate it simply by covering with cardboard.

The advantage of a living mulch is that, while the roots bind and feed the soil, the dead leaves add to organic matter without any disturbance. However, you do have to watch the relationship between the crop and the mulch. The living mulch needs to be cut back if it gets too tall and starts competing with the crop for light, water and nutrients. I have tried red clover (*Trifolium incarnatum*) under squash, but the plants were too competitive, so I use white clover. You can even use weeds such as ground ivy (*Glechoma hederacea*) or chickweed (*Stellaria media*) as a living mulch, while tomatoes can be undersown with coriander (*Coriandrum sativum*) or summer purslane (*Portulaca oleracea*).

Another way to use a living mulch is to intercrop by establishing strips of the clover between rows of the crop. This method works best in larger spaces, such as an allotment or smallholding, where you can grow rows of brassicas, such as Brussels sprouts, or pumpkins and squash. Establish the strips of clover, and then plant your crop in between. You may need to mow regularly to keep the living mulch from becoming too competitive. Buckwheat and annual rye can also be used in this way.

The value of leaf litter

For me, leaf litter is invaluable for boosting soil health. Not only is there plenty of it, it's also free and easy to compost. I collect fallen leaves from paths and driveways, and either pop them in a one-tonne bag or a mesh-sided bin and leave them to rot down to leaf mould, which takes about a year. The resulting mould can be used as a mulch or as an ingredient in your own peat-free potting compost.

Once I have collected all the leaves I need, I sweep those lying on paths onto my flower beds. Some gardeners claim that you should clear them up. Are they right? I always look to nature and I find that soil and plants seem to cope quite well with fallen leaves. So, a few points to consider:

Regenerating your soil

- Fallen leaves create a mulch that protects the soil from winter weather and suppresses weeds.
- Fallen leaves complete the cycle of nutrients, but remember that beech and oak leaves take much longer to break down than those of birch, lime, hornbeam and hazel.
- Some gardeners say that it's important to remove fallen leaves to let the soil breathe. This is incorrect. The soil 'breathes' perfectly well, and the leaf litter protects the soil from the elements and stops nutrients washing out, while the leaves feed the earthworms that tunnel through the soil and improve air circulation and drainage.
- Leaf litter creates the perfect habitat for many beneficial animals, such as ground beetles and centipedes.
- If you have autumn bulbs or low-growing plants, make sure they are not buried under a thick layer of leaf litter, as it can trap too much moisture and encourage rot. Don't let leaves pile up around trunks and stems, as they can be a source of rot. A thick layer can also make the perfect habitat for voles, which like to gnaw at wood over winter.
- You may have read advice about removing leaves from lawns. You don't want to leave your lawn completely covered in a deep mulch of leaves, as it will kill off some of the grass, so the best method is to run a mulching mower over them so they are shredded and break down very quickly.

Should you clear up diseased leaves and burn them? This is classic advice for fungal diseases such as tar spot on acers and sycamore, but this disease doesn't really harm the tree; it just looks unsightly. Even if you clear up the leaves under an infected tree, will you catch them all? What about the ones that have blown away? Furthermore, it seems that sweeping up these leaves barely has any effect on the level of tar spots the following year. Another disease is apple scab. Again, the advice is to collect all the leaves and dispose of them, as the fungus overwinters on fallen leaves and in the soil, so the theory is that removing them lessens the number of spores in spring. But it's far better to run a mower over the leaves so they are shredded and break down quickly. If you are concerned, remove the diseased leaves and compost them but, more importantly, look at why your apples are getting the disease and work on boosting soil health around the diseased trees.

No-work, deep mulch

I love this variant of no-dig gardening, which I first discovered when researching no dig some 16 years ago. American gardener Ruth Stout was born in Kansas in 1884 and had a novel approach to growing vegetables. She called it 'no-work, deep mulch'. It involved keeping a thick mulch (20cm/8in) of organic matter on her vegetable and flower beds all year round. She added more during the year to cover up any weeds that appeared and sowed and planted through the mulch, moving the mulch back to drop in the seeds before covering it again. I've seen this method used successfully on an allotment, too. The allotmenteer spread a thick layer of straw over his plot and visited just a handful of times – to plant out runner beans, squash and corn, carry out checks and, finally, to harvest! Nowadays, I use a similar method on my brassica and squash beds – covering the soil with a thick mulch of hay in autumn and then planting in mid-spring. Under the thick mulch, the soil stays moist all year and I don't have to water. To reduce the risk of slugs, I spray with beneficial nematodes before planting out and again six weeks later (see 'Table 4. Beneficial nematodes', page 122).

WOODCHIP

A woodchip mulch can boost the establishment and growth of perennial trees and shrubs by trapping moisture, suppressing weeds and encouraging fungi, all of which lead to a healthy soil. It

Regenerating your soil

works to loosen compacted soil, too. In many cases, it's better to add woodchip rather than compost, as compost has the potential to add too many nutrients to the soil and encourage lush plant growth, which is more prone to pests and disease.

I'm a great fan of woodchip for mulching, especially willow woodchip. I use much more than I can currently generate myself, so I ask a local tree surgeon to drop off a few loads, although I find I am not the only one asking for this material now! I don't have much say in what arrives and, quite often, it is conifer-based while, ideally, you want a mix of different species with no more than about 20-30 per cent conifer.

If you find that you have a pile of conifer chips, don't despair; just let it age. Conifers release chemicals that inhibit the growth of deciduous species. This is called allelopathy (see 'Root wars', page 140). The best use for fresh conifer chips is to spread them on paths to suppress weeds. If you let the pile rot down, the brown rot fungi will get to work and, eventually, all the chemicals will be broken down and you can use the mulch safely on your beds. But even then, I would mix it with deciduous woodchip to be safe.

When ordering a load, remind the tree surgeon that you don't want woodchip that contains any diseased materials. You certainly don't want to bring in disease, especially honey fungus, but a good tree surgeon should know this.

Younger woodchip

There is one form of woodchip that holds even greater promise than standard woodchip. It is known as ramial chipped wood (RCW). This is woodchip made from deciduous branches less than 7cm (2¾in) in diameter. You can get ramial woodchip from the prunings of trees and shrubs. The reason why this material is so valuable is that it is young wood, rich in nutrients and low in lignin, which means it can be used fresh. In fact, it's estimated that 75 per cent of the minerals, amino acids and proteins in a tree are locked up in its young branches. Such branches have a Carbon:Nitrogen ratio of between 30:1 to 170:1, whereas mature wood is 400:1 or more. This makes a huge difference to the availability of nitrogen, as it doesn't take so much decomposition to make it accessible. Ramial woodchip can be broken down by white rots while mature woodchip, with its higher lignin levels, is broken down by brown rots in a much slower process.

If you haven't got time to chip all your winter prunings, cut them into shorter lengths and drop them on the ground around your trees where they will rot down. Permaculturalists call this method 'chop and drop'. But always take away diseased material, especially if a plant is suffering from canker.

Woodchip destined for paths can go on straightaway, but I don't put fresh woodchip around plants. Instead, I leave the woodchip pile to rot down and, once it's shrunk and I see white fungal hyphae spreading through the pile, I use it as a mulch around fruit trees and bushes and perennial plants, and I'll even spread some over my vegetable beds.

Grow your own woodchip

One completely sustainable option, if you have the space, is to establish some fast-growing plants to produce your own woodchip. Willow is a great choice, as it is easy to establish from cuttings

A word of warning

I am careful to wear a good-quality builder's face mask when moving woodchip, as I have learned from experience that inhaling fungal spores is not good for the lungs. In fact, there have been cases of gardeners getting aspergillosis, so I don't take any risks.

and can be coppiced (cut to the ground) every two to three years. I find it easiest to lay down a strip of groundcover fabric or cardboard to keep the weeds down and push short lengths of willow stem through this into the ground to half their length. Watch out for rabbits, though, as they are partial to nibbling the young shoots. Let the plants establish for a couple of years and then cut and chip. You can establish a row of willow as a boundary hedge. Not only does it look attractive, but it also brings in wildlife and creates windbreaks and screens. If you want willow for both a screen and a supply of woodchip, establish a double row of plants and alternate which row you cut. Another option is hazel, which is also good for wildlife.

HOW HEALTHY IS THE SOIL IN A RAISED BED?

Raised beds are popular with gardeners and I see endless photos of garden bloggers on Instagram showing off their newly designed vegetable areas with rows of neat, raised beds, many formed using wood planks. Although the vegetable beds in my allotment area are traditional in style, I have some raised beds in my formal walled garden. Their soils are well drained, warm up quickly in spring and don't suffer from compaction. They are great for Mediterranean plants that need free-draining soils which don't become waterlogged in winter. But they can require more watering in summer, as the soil can dry out more quickly.

On the downside, raised beds can be costly to build and fill, as you often have to buy in soil or compost, and they require a surprisingly large volume of soil to fill. Another disadvantage of raised beds, especially those made from wood, is that they provide hiding places for pests, such as slugs and snails. I frequently find their eggs tucked down the inner sides of the wood planks, especially those that have been in use for a few years and are starting to rot. Also, I have found that our local vole population loves tunnelling along the inside of the boards.

Building a healthy soil

But how healthy is the soil? The more I learn about the rhizosphere, the more I am tending to move away from raised beds to traditional beds or even hügelkultur beds. One thing I do like about a traditional bed is the lack of barrier between path and bed, so the plant roots and fungal hyphae can extend through the soil under the path. It's the perfect transition zone where two habitats overlap, so there will be more diversity at this point in the garden.

I'm a great fan of hügelkultur beds. I first read about these in the book *Sepp Holzer's Permaculture*. These beds take some effort to establish because you have to dig out a trench which you backfill with old logs, leaf litter and even cardboard and straw, to create a long-term source of carbon. You then cover the bed with smaller branches and woodchip and top it off with either downward-facing grass sods or compost. The logs create gaps so there is plenty of aeration, water storage, great drainage and a reservoir of nutrients. My own hügelkultur bed has performed really well, surviving the 2018 drought year with no watering. It sinks slowly as the material rots down, so I add more woodchip on top to replenish, and that's about it.

Despite my reservations, I'm still using raised beds in my kitchen garden, but now I make sure there is a layer of old logs and leaf litter at the bottom, hügelkultur-style, to encourage the fungal community in the soil.

PART 2 **Pests and predators**

CHAPTER 3

Understanding pests and diseases

'If something is not eating your plants, then your garden is not part of the ecosystem.' I don't know who to credit for this brilliant phrase, but they understood that if you have a healthy, biodiverse garden, there is always something eating something else and that, should pests arrive, there will be plenty of natural predators around to keep them in check. Knowledge of the life cycle of a pest or a disease-causing organism helps you to keep control, prevent it from thriving by not providing it with food, taking away its shelter and overwinter places, and boosting its predators. In this chapter, I look at some of the pests and pathogens that we may find in our growing spaces and how they interact with our plants, plus I take a look at garden hygiene.

PLANT PESTS

Plant-eating insects make up half of all the known insect species and, not surprisingly, are common pests in our gardens. They are categorised according to the different ways they attack plants:

Pests and predators

Ants and aphids

Many insects have a close relationship with ants, so watching out for ants is a good way of keeping track of pests. Ants and aphids have a mutually beneficial relationship, so seeing ants running up and down plant stems is often an early warning of trouble ahead. Sap suckers cannot use all the nutrients they remove from the plant, so the excess is excreted from their body as honeydew. Honeydew is rich in sugar and is a favoured food of ants. Ants milk aphids for honeydew by stroking their abdomen and, in return for this valuable food source, they protect the aphids from predators. If you disturb aphids, you'll probably see ants rushing over to protect their charges. To get the upper hand, you need to control the ants. You can wrap the base of the stem in a sticky tape to catch them and prevent them looking after the aphids. Once the ants are out of action, predators will be able to feed on the aphids.

Sap suckers Including aphids, scale insects, thrips and whitefly, these are the commonest of the insect pests. They have specially adapted mouth parts called stylets. These long thin tubes can pierce a plant stem and locate the tubes that carry sugar-rich sap around the plant. Sap suckers tend to weaken their hosts rather than kill them, so the plants can recover from an

Understanding pests and diseases

infestation if they are growing in healthy soil and are not stressed by other factors, such as drought or waterlogging. Another risk is that these insects can carry pathogenic bacteria and viruses between plants, which may prove more deadly than the sap sucker itself.

Borers These insects spend much of their life inside plants, tunnelling through the stems of trees, shrubs and crops. Most boring insects are beetles, moths and wasps and, very often, it's the larvae that do the damage. The adults lay eggs on or near the host plant and the larvae enter the plant to feed. Some pupate within the plant, while others drop to the ground where they pupate in the leaf litter. The adults emerge and the cycle begins again.

Defoliators These are insects with chewing mouth parts, such as caterpillars, grasshoppers, leaf miners and webworms. They can strip a plant bare of leaves, leaving it in a stressed condition and susceptible to disease.

Deflowering insects These remove the flowers of a plant, reducing its potential to produce fruits and seeds. The first signs are holes in buds, buds that fail to develop into flowers and buds that drop off. These pests live inside the bud, eating the flowers from the inside out. They include moth caterpillars that eat flower buds on ornamental plants.

Gall-making insects These live inside a leaf, stem or twig, where they cause the tissue to swell, creating the gall structure. These are odd-shaped lumps that are made of plant tissues but caused by another organism. Galls don't tend to harm the plant, although they can look unsightly and disfigure the leaf or stem. In the case of gall wasps, the female wasp lays an egg inside the plant tissue. The larva spends its life inside the gall, then pupates and the adult wasp exits the gall.

Slugs and snails

Ask any gardener to name their top garden pest and I suspect it will be slugs. These pesky molluscs plague gardens as far afield as New Zealand, the USA and across Europe, as well as in the UK, munching through newly planted crops, leaving holes in leaves and glistening slime trails across the garden. I get plenty, but I don't want to wipe them out completely as they have a role to play in food chains.

Slugs use their rasping tongue to eat mostly decaying leaves, damaged fruit and even animal droppings and dead bodies. It's claimed an average slug can eat 40 times its weight in a year, but only a small proportion of this is living plant material. There are more than 40 species of slugs in the UK, but only a couple are considered pests. You may not like the look of the large fat slugs, but they generally prefer rotting plant material and dead bodies rather than your precious plants. The ones to watch out for are the grey field slug (*Deroceras reticulatum*), which is found worldwide, and the various keeled slugs.

Given the ubiquitous nature of the problem, I thought I would show how it is possible to tackle a problem pest through natural approaches, making use of our knowledge of their life cycle and ecology to keep them under control. Some of these ideas are more effective than others, and ideally, you need to use a

Understanding pests and diseases

combination of methods to keep their numbers at an acceptable level.

Encourage plant diversity Slugs tend to do less damage in diverse systems because they have to keep switching to different plants and so eat less in total, whereas they can feed more effectively when they find a large area of a single crop, so polyculture is better.

Encourage earthworms A healthy soil has plenty of earthworms and an Austrian study from 2013 showed that the presence of earthworms protected the plants and reduced slug damage by as much as 60 per cent.

Remove hiding places Slugs like to crawl under rocks, logs, pots and garden furniture, etc, so remove these and keep your pots off the ground with footers. They love rotting wood, so you'll find them hiding down the sides of old raised beds made with wooden planks where they lay their round white eggs. If, like me, you encourage wildlife by leaving fallen leaves on the ground and building log piles, for example, you end up providing the ideal places for slugs as well as the predators, so try to place these features as far away from your vegetable beds as possible.

Use plant deterrents Plants such as astrantia, wormwood, rue, fennel, mint and rosemary have been found to act as a deterrent. It's

Pests and predators

thought that they give off a scent that repels slugs.

Encourage natural predators Blackbirds, starlings, frogs and toads, slow worms, hedgehogs, ground beetles, centipedes and rove beetles all feed on slugs and snails.

Move the compost heap You want to encourage slugs and snails near your compost heap but away from your growing areas.

Cut back low-hanging branches Those dragging on the ground create dark, damp hiding places for slugs and snails.

Cut grass regularly Keeping the grass low near vegetable beds and vulnerable plants will deter slugs and snails.

Hoe beds and borders Hoeing creates a friable soil texture that dries out quickly, resulting in a rough surface which is less appealing to slugs.

Harden off vulnerable plants Move plants raised in a greenhouse or polytunnel into a cold frame to thicken their leaves and put on a bit more growth. If they have a larger leaf surface, any slug damage will have less impact and the plant can recover. Avoid using too much nitrogen fertiliser as this encourages lush, susceptible growth.

Delay sowing vulnerable crops If the weather is cold and wet, delay sowing crops like carrots. You want the seedlings to germinate and get away quickly. Delaying for

Understanding pests and diseases

a couple of weeks until it is drier and warmer could make all the difference.

Mix the crops Don't plant vulnerable crops in blocks, but instead mix them with crops that slugs don't like or plant a barrier of nasturtiums. You can do the same with ornamentals: slugs and snails tend not to like grey foliage or downy, waxy or thorny leaves.

Water wisely Water in the morning, so the soil dries out before the slugs and snails are active.

Crop choice If you suffer slug damage on potatoes, go for varieties that are less susceptible.

Mulch at the right time Although good for conserving moisture, mulching can also encourage slugs, so if the spring is cold and wet, don't mulch too soon; otherwise, you are just creating the perfect hiding places for slugs.

Control methods

The RHS studied five popular methods of controlling slugs and snails – copper tape, sharp grit, pine bark, wool pellets and eggshells – using lettuce grown in pots and in raised beds for six weeks. Plants in the ground were found to be more susceptible to slug and snail damage than those in pots, but none of the treatments warded off slugs and snails. The layers of wool and pine bark were found to be beneficial as a natural fertiliser and mulch, and the plants were 50 per cent larger, but the numbers of slugs were the same. It was concluded that the molluscs had such thick mucus covering over their foot that they were unaffected by the different surfaces, although they preferred a moist, organic mulch over a dry, gritty one. Bearing this in mind, here are some suggestions for controlling slugs and snails.

Collecting Slugs and snails are nocturnal, so go out on a damp, warm night and collect them. Search under leaves and logs, and look for their slime trails, which glisten in torchlight. Picking slugs up by hand is messy because they're slimy, so some people use chopsticks or spear them on a stick or knife. Once caught, you can move them to the compost heap or drown them, but I find it easiest to cut slugs in half with my penknife. Snails hibernate in winter, crawling into sheltered spots where they seal their shells and become inactive, so it's worth hunting them out.

Traps An upturned grapefruit skin or wilted piece of cabbage leaf will attract slugs. Place around vegetable beds and go out at night to check your traps.

Beer traps It's the gas from the fermenting yeasts in beer that attracts slugs. The slugs crawl into the trap and die, and their dead bodies add to the fermenting mess. The traps

Understanding pests and diseases

only work over a short distance and beers differ in their effectiveness! It also doesn't have to be beer, as the key is yeast, so sugar and baking yeast is just as effective.

Trap plants Grow trap plants to draw in the slugs, leaving other vegetables alone. These sacrificial plants can be planted between or around those you want to protect. Try basil, lettuce, radish and spinach and, for the ornamentals, you can plant asters, dahlias and petunias. Chamomile has been found to attract slugs, too. It starts growing in early spring, just when the slugs and snails are getting active. Rubbing the surface of a patch of chamomile is said to attract nearby slugs and then you can pick them off.

Beneficial nematodes This is one of the most effective treatments (see 'Table 4. Beneficial nematodes', page 122). It's relatively expensive and needs to be applied several times through the growing season, but if you use this treatment over a number of years, the numbers of slugs and snails will decrease.

Ducks They love slugs and snails. When I let my ducks out in the morning, they patrol their pens, scooping up all the slugs they can find. Let them into your growing areas after harvest to clear up eggs or before you plant out in spring. Ducks are ideal for clearing raised beds as they push their beaks down the sides looking for slugs, snails and egg

Pests and predators

clusters. The call duck is great as it's small and doesn't do much damage to beds, so can be kept permanently in the garden or growing area.

Coffee grounds Although not an effective barrier, research found that a 1–2 per cent solution of caffeine kills slugs and snails and was more effective than the toxic slug killer metaldehyde. Although instant and filter coffee contain less caffeine, pouring cold coffee on your worst-affected beds could act as a mild deterrent. But remember that coffee is very acidic and will alter soil pH if used regularly (this is fine for ericaceous plants). Plus, it is not selective and so will harm other invertebrates.

Slug pellets Organic slug pellets are based on iron phosphate, not metaldehyde, and are approved for organic gardening, but they are not completely non-toxic to other animals and can kill earthworms if ingested. Also, these pellets have other active ingredients such as EDTA (Ethylenediaminetetraacetic acid) to make the iron phosphate more attractive to slugs and snails, but these may also make it more attractive to earthworms and pet dogs.

And finally, some questions to think about. I have had plenty of discussions about slug control with fellow gardeners on social media. Is there

Understanding pests and diseases

> more of a slug problem in an urban garden than a rural one? Does a small urban garden with less habitat diversity lead to more slugs? Are there fewer birds to control the slugs? If your neighbour uses chemicals, does this affect the number of natural predators in neighbouring gardens? I don't have the answers, but these are things to consider.

PLANT PATHOGENS

Plants are affected by a wide range of pathogenic bacteria, viruses, fungi and nematodes, and their health – particularly their nutritional status – can affect their susceptibility to these pathogens. Often, it's a deficiency in micronutrients such as copper, iron, silicon and zinc that results in the plant succumbing to a pathogen.

Plant pathologists describe the relationship between a plant, the pathogen and its environment as a triangle. You need all three to get disease. The pathogen must be able to come into contact with the host, gain entry via a wound, natural opening or through direct penetration, then establish itself within the host, grow and reproduce on the host, and then spread to other susceptible hosts in order to ensure its survival. This is the disease cycle and, to prevent disease or reduce its effect, you need to disrupt this cycle.

Watching for symptoms in infected plants will help you to take action quickly. For example, look out for yellowing, browning, mottled or wilted leaves. Also watch for dead tissues on leaves, stems and roots that appear first as spots but then spread to form larger areas. As the disease progresses, the plant may drop its leaves and fruits, leaves may curl, and growth may become stunted.

Pathogenic bacteria

A huge diversity of bacteria cause plant disease, such as blights, cankers, galls, scabs and wilts. Bacteria are spread by rain and wind or are carried between plants by insects and birds; they can also be transferred by tools and shoes. Entry is gained via stomata (pores) on the leaves and other openings, or through wounds. Once inside the plant, the bacteria may produce toxins or inject proteins that kill cells or produce enzymes that attack cell walls. Soft rot bacteria, for example, break down the layer of pectin that holds plant cells together. This causes the tissues to lose structure, collapse and die. Bacteria that cause wilts invade and block xylem vessels, which carry water, stopping the movement of water from the roots to the leaves, so the leaves wilt and die. Bacterial diseases can be difficult to control but good hygiene in the garden can help reduce bacterial spread.

Viral diseases

Viruses need a living plant to infect. They have no means of forcing an entry into a plant, so they either need a wound or a vector such as a sap-sucking insect. Once viruses are inside the plant, they take over the host's cells to make multiple copies of themselves which infect more cells and cause symptoms of disease to appear. Once infected, it is impossible to control the virus, so the plant is usually destroyed. Just as with bacterial diseases, the best way to prevent viral disease is through good hygiene. Some viruses only cause minor effects or remain symptomless. And there are a few that actually cause aesthetically pleasing changes in the colour of flowers or leaves. The classic example is colour break in tulips, where the presence of a virus results in variegated petals because the pigment is not uniformly distributed. Other examples include geraniums, impatiens and camellia. However, in most cases, the flower may look attractive, but the plant itself won't thrive.

Pathogenic fungi

Fungi are key members of the soil food web, so it's confusing when many of the common diseases of plants are caused by fungi and

Understanding pests and diseases

fungi-like organisms. They cause more plant diseases than bacteria and viruses, and almost all spend part of their life cycle on their host plants and part in the soil or leaf litter. Historically, fungi have played a crucial role in history, contributing to famine and economic losses: bunt destroyed wheat crops in the Middle Ages, potato blight (*Phytophthora infestans*) destroyed potato crops in Ireland, causing famine and mass migration, while vineyards across Europe were hit by downy mildew in the 1870s.

Several groups of fungi cause disease in the garden, along with a fungi-like group called Oomycetes, the water moulds, which include damping off (*Pythium*) and blight (*Phytophthora*), two diseases that thrive in damp conditions. Most fungi and Oomycetes produce spores which spread the disease, but there are some that lack spores, namely the *Rhizoctonia* that cause root diseases in many crops.

Fungal pathogens fall into three main types, based on their relationship with the host plant:

Biotrophs These are highly sophisticated pathogens that grow within living plant cells without setting off the plant's defences, remaining 'hidden' and spreading quickly. Downy mildews, rusts and smuts are typical examples. They can only survive on their host, so don't tend to kill it. Spores remain dormant on the ground until they are picked up and carried to a suitable host plant. When it comes to disease control, remember that these fungi cannot survive without a host, so remove infected plants.

Semi-biotrophs Spending most of their life cycle on the host plant, these pathogens switch to feeding saprophytically on its dead remains once it dies. However, they do not feed on other plants, dead or alive. They include apple scab (*Venturia*), blight (*Phytophthora*), moulds (*Botrytis*) and powdery mildews. It is important to clear away infected material and so remove a potential source of food for the pathogen.

Pests and predators

Necrotrophs This group kills their host by secreting toxins and, once the plant is dead, they feed as saprophytes. *Pythium* is a necrotroph that causes damping off in seedlings. Most necrotrophs feed on all available food sources, not just that of their dead host. They tend to be soil-based and infect a wide range of hosts. They can remain dormant in the soil, surviving for many years.

Fungal spores are distributed on the wind, by rain splash or are carried by insects, birds and even on the fur of mammals. Eventually, a spore lands on the leaf of a host plant where it germinates and produces a hypha that enters the plant. Some fungi, such as the saprophytic sooty moulds that feed on honeydew, stay on the leaf surface where they form a black layer. Those pathogenic fungi that infect roots are found in the soil and their spores are often motile (able to move) and can make their way to the roots via the film of water around soil particles.

There is a wide range of pathogenic fungi that can affect crops. The main groups include:

Brown rots These are a bit of a 'Jekyll and Hyde' group of fungi, as the action of brown rots in decomposition and composting is crucial to healthy soil while, at the same time, they are probably the most common disease affecting the blossom and fruits of fruit trees. Typically, these fungi invade flowers and fruits and turn them into a mush, with grey-white pustules on the surface that release spores.

Mildews Both downy and powdery mildew infect the aboveground parts of plants, especially leaves, but they rarely kill their host. Their white, fluffy growth on leaves and stems is produced by masses of spore-producing structures.

Spots These fungi produce spots on leaves or fruit and, while they weaken the plant, they rarely kill it. Generally, each fungal species infects only one host species. If the spots become

Understanding pests and diseases

large and join up to cover a large area, the disease is classed as a blight.

Rust and smut These fungi erupt through the plant's epidermis to produce masses of coloured pustules on the leaf surface. Rust fungi produce rust-coloured spores and smuts produce masses of black spores.

Wilts The hyphae of fungal wilts grow into and block the plant's xylem vessels, causing leaves to wilt. Some gardeners mistake leaf wilt as a water shortage, but it is often a result of root damage caused by the fungus.

PARASITIC NEMATODES

Nematodes are microscopic, tube-like animals found in soil. Some are parasitic, some decomposers and others are beneficial predators. The plant parasites have piercing mouth parts to penetrate and gain entry to all parts of the plant. Most common are the nematodes that attack roots, causing root knots, lesions, wilting and stunting. Less common are nematodes that damage leaves, buds and bulbs. Nematodes are spread by equipment, shoes, wind, water and animals and also by infected plant material. Control is achieved through hygiene and management (see 'Garden management and hygiene', page 61), rotating crops (see '1. Crop rotations', page 93) and using cover crops to repel nematodes (see 'Biofumigation', page 141). There is also a group of predatory nematodes that prey upon plant pests, such as slugs, which are used as biocontrols (see 'Nematodes', page 119).

NEW PESTS, MORE PESTS

As climate change makes itself felt, we may see changes in the threats from pests and disease: some disappearing and new ones appearing. Warmer summer temperatures could mean that pests complete their life cycle more quickly, so there are more

Pests and predators

generations and an increased risk of epidemics. For example, the long, dry summer of 2018 saw high numbers of flea beetle, dock beetle and defoliating micromoths.

Some insects can't survive the winter as adults, so they overwinter as an egg or larva. However, milder winters could see adults overwintering, too, and even becoming active during sunny spells. Overwintering adults would mean an earlier start to breeding. Many pests are seen only in the summer because they are migrants from Europe, but warmer winters could lead to more surviving and becoming permanent residents. Milder springs would see pests getting an earlier start. Scientists have calculated that for every 1°C (34°F) increase in average temperature, aphids will become active two weeks earlier. We may even see some greenhouse pests, such as red spider mite, surviving outside and damaging garden crops.

Some diseases could become much more common, especially fungal and bacterial pathogens such as *Phythopthora* and *Pythium*. A rise in blackspot may make growing roses trickier, while warm, dry summer weather will help the spread of powdery mildew.

Extreme weather stresses plants, especially trees, making them more susceptible to disease. Warmer, but wetter, winters could see more fungal diseases, such as honey fungus and apple cankers. New diseases are already appearing and getting a foothold, while others which were relatively rare are now more common. For example, box blight (*Cylindrocladium buxicola*) appeared during the 1990s in Hampshire, but it is more frequent nowadays and its range is extending northwards. Similarly, the spread of holly leaf blight (*Phytophthora ilicis*), first seen in Britain in the 1980s, is helped by warmer, wetter winters. Olive scab (*Venturia oleaginea*) is a new arrival that was imported on infected plants and has the potential to infect newly planted olive trees. Yet another threat is *Xylella fastidiosa*, which is already spreading through olive groves in Europe. There are fears that it could be imported into the UK on olives and other host plants, such as lavender.

Understanding pests and diseases

It's possible that tropical diseases may get a foothold, too. One that the experts are watching is *Athelia rolfsii*, a soil-borne fungal disease called southern blight or sclerotium rot, which is widespread in tropical and subtropical areas where it infects a range of species, including tomatoes. It is found in Europe and, if introduced into the UK, would find plenty of plants to infect, especially if it could survive the winter. In North America, fire blight, caused by the bacterium *Erwinia*, was once restricted to more humid regions in the south, but climate change is seeing it become more of a problem in the north.

GARDEN MANAGEMENT AND HYGIENE

There are simple things a gardener can do to keep disease in check. For example, fungal diseases thrive in humid environments. Close planting encourages humidity, so you can encourage better air movement by trimming plants that are dragging on the ground and by staking others, so their stems are off the ground. If you can, train a plant up a trellis or frame rather than letting it scramble over the ground. Also, try to avoid working around plants suffering from fungal disease during wet weather, as the spores are easily dislodged and will be carried on clothes and boots.

Fungal spores are carried long distances on the wind, so a windbreak can help to limit their spread. In some agroforestry systems, a row of trees beside a crop of potatoes has been found to reduce the incidence of potato blight, so growing your potatoes downwind of a hedge, row of willow or even a row of Jerusalem artichokes can slow down the spread of the spores. I'm a bit neurotic about spores of late blight being carried into my polytunnel in late summer, just when the tomatoes are in full production. I don't grow potatoes near the polytunnel and, if I have been working around potatoes with blight, I clean my shoes or wellies with a disinfectant that is effective against bacteria, viruses and fungi. It's not just blight, either – there are a number of soilborne diseases that can be transferred by boots or muddy tools, including club roots, mildews and onion rots.

Bacteria can be transferred by tools as well. I am careful when pruning my fruit trees, as I have had canker in the orchard. Because I don't want to transfer bacteria from an infected tree to a healthy one, I now wipe my secateurs and pruners with disinfectant between trees.

What about sterilising pots and containers?

A quick trawl on the internet and you will find endless advice about scrubbing your pots and containers with anything from 10 per cent bleach to proprietary disinfectants. This is something I don't do and have never done. I have a huge pile of plastic and clay pots, all sorted into sizes in the shed, which I just grab and use. If a pot is dirty, it might get a dunk in water, but that's all. The wisdom behind pot sterilisation is to reduce the risk of introducing spores that cause damping off in seedlings but, if you think logically about this, anything lying around, such as tools, a garden hose or gardening gloves, can carry spores, as can your hands. I think the risk is more likely to come from a poor-quality compost and seeds and less-than-optimum growing conditions (too cold, too wet, not enough light, etc.) than reusing an old pot or tray.

I don't sterilise my polytunnel at the end of the season either, partly because I use the space all year round and never have an 'end of the season', and partly because I am wary of the chemicals that are recommended, especially sulphur candles. These are lit to fumigate the space and kill pests, but the toxic fumes are not pest-specific, so you are just as likely to kill off your overwintering predators and other beneficial animals.

What to do with infected garden material?

Just like cleaning pots, there is much debate about what to do with infected material. I definitely think it's important to get rid of diseased plant debris infected by necrotrophic fungi that can survive on dead and decaying plant material. This also stops pests, such as

Understanding pests and diseases

aphids, carrying the disease to more plants. Ideally, you should hot-compost the remains.

Many pathogenic fungi are biotrophs, so require living material and won't survive in the ground. While many say it's important to burn blighted potatoes and tomatoes, the spores don't survive long on the ground and they won't survive on the compost heap either, especially a hot one. If you have had a problem with a pest or disease, check its life cycle to see if it will survive in the ground over winter. If it does, then you need to make sure you rotate your crops onto fresh soil to avoid more problems. For example, cabbage and carrot root flies can survive in the soil for several years, so don't grow the crop on the same bed if you have had problems.

It's also important not to bring pests and disease into your garden. I've lost count of the number of gardeners who have told me that they have imported a problem, especially the vine weevil, in potted greenhouse plants. Recently the UK and EU regulations regarding plant health have been overhauled to prevent the introduction of non-native pests. Now, people who move plants and plant products from business to business within the UK and EU have to make sure that, where required, the plants are accompanied by a plant passport label showing the country of origin of the plant and a traceability code.

KEEP ALERT

The key to keeping pests and disease under control is observation. Look at your plants regularly, watch out for white butterflies that will lay eggs on your brassicas, check closely for aphids and ants, look for clusters of eggs under leaves and other tell-tale signs. Note-taking is really useful as it helps you build up a record of when a pest or disease was first seen, which plants or beds were affected and how it spread. It doesn't matter if the notes are handwritten or photographic, it's the records that are important, so you can look back over your records and assess the risky periods.

Pests and predators

One of the difficulties gardeners experience when switching from a traditional chemical approach to pest and disease control to a natural one is the risk of outbreaks and flare-ups as the system is settling in. So, you have to be observant, act quickly when you see pests, and be prepared for an upsurge in numbers. Get good management practices in place and make sure your garden environment favours natural predators, which is covered in the next chapter.

CHAPTER 4

Natural predators

Our gardens and allotments are home to many natural predators, from the familiar ladybirds and spiders to the less familiar rove beetles and parasitic wasps. These natural predators, or 'beneficials' as they are called, are your first line of defence.

Natural predators are important in all habitats, natural and artificial. In nature, the populations of wasps, ladybirds, spiders and birds have long regulated the abundance of pests. For example, in a stable forest ecosystem, predators keep 90 per cent of the arthropod herbivores below outbreak levels. Spiders control Douglas fir tussock moth caterpillars and birds will prey on gypsy moths. Under normal conditions, predators can prevent outbreaks when pest numbers are low to moderate, but problems can arise when there is a change in the weather or a natural disaster, such as a fire, flood or drought, which causes the balance between the prey and predator to be disrupted.

THE PREDATOR-PEST BALANCE

Much research has been carried out on predator-pest relationships and it is clear that systems with a low biodiversity are particularly vulnerable to pest outbreaks. This is what we must try to avoid in our growing spaces. In my own garden, I am aiming for a mosaic of habitats in order to attract a wide range of beneficials

Pests and predators

to prey on different types of pests. Over the last 10 years, I have seen a steady increase in biodiversity and now it's not unusual for me to wander around the garden and spot a variety of predators. I still have pests, of course. In order for these predators to thrive, there must be some prey animals, so I have had to learn to live with low levels of pests to keep my beneficials fed! A key challenge, though, is to make sure that there are enough predators around early in the growing season to control the problem aphids, whitefly, red spider mite and various larvae, so that their numbers don't spiral out of control. The peak season, as far as pests in the garden are concerned, is early summer, so you need the beneficials to multiply rapidly to take advantage of the food bonanza, and this needs to continue through mid-summer. You can achieve this by understanding their life cycles and dietary requirements as well as providing the right plants and overwintering habitats to enable them to thrive. Clearly, it's also very important to be able to identify predators at each stage of their life cycle.

Don't overreact

It's important not to overreact when you see pests. For example, you might notice that your broad beans are smothered by blackfly and decide that you need to do something quickly to prevent serious damage to the crop. But this is where you have to be careful. If there are lots of prey around, it is likely that you already have increasing numbers of predators, as the predators will lag behind the prey in the population stakes. If you are tempted to 'take the nuclear option' and use a chemical spray, you may well take out both prey and predator and, in doing so, you will destroy the natural cycle and won't then have any chance of being able to get back some natural control. It is best, if you can, to be patient, let nature take its course, and wait for the predators to 'kick in'. If you can't wait, check carefully for any predators on the plants before using an organic product or simply use your fingers to remove or squash the pests.

Natural predators

GARDEN PREDATORS

You can expect to find a wide range of beneficial animals in your garden or allotment, including ladybirds, lacewings, hoverflies, ground beetles, parasitic wasps, braconid wasps, tachinid flies, social wasps, spiders and centipedes. Among the vertebrates, there are frogs and toads, lizards and slow worms, birds and mammals. You can enhance the population of these predators by growing the appropriate plants and providing overwintering places. Here's a quick overview of the main groups of garden predators:

Beetles

Most people probably recoil when they lift a pot and a large black beetle rushes off into the shadows, but I love to see them thrive in the garden and polytunnel. There are various beetles in the garden, including ground beetles, ladybirds and rove beetles. They are all ferocious predators and important allies in the garden. Ground beetles chase down their prey on their long legs and grab them with their powerful jaws. For their size, they are among the fastest creatures in the animal kingdom. These predatory beetles are relatively long-lived (up to four years or more) and can reach a length of 30–40mm (1¼–1½in). Most are nocturnal, hiding under leaf litter, logs and pots during the day and hunting at night for a variety of animals, including slugs, snails, caterpillars, leatherjackets and aphids. They lay eggs in the soil or leaf litter and their soil-dwelling larvae are large, fat and brown-black in colour with three pairs of legs. They feed on soft-bodied animals and eggs, especially snail eggs.

Rove beetles or 'staphs' belong to the Staphylinidae family. They make up a quarter of all British beetles and are often mistaken for earwigs. They live among leaf litter and in compost heaps. Fast and agile, they have a flexible abdomen that enables them to chase prey through tight spaces.

Pests and predators

Ladybirds

Also called ladybugs or ladybird beetles, ladybirds are easy to identify. Most are easily recognised by their red or orange colour with black markings, although some are black with red spots and others lack spots. They are voracious predators, catching and chewing their prey. Ladybirds can live for up to three years, overwintering in clusters under leaf litter, bark and rocks and also in buildings and sheds. They emerge in spring to feed on the pollen and nectar of early flowers but, as soon as their prey becomes active, they move to gardens and vegetable plots to lay eggs. It's here that they spend the summer, feeding and laying more eggs. Both adults and larvae are predatory, and their preferred food is aphids. It's estimated that a single ladybird can eat more than 5,000 aphids in its lifetime. If aphids are in short supply, ladybirds will feed on pollen and nectar and other prey, including mites, thrips and all sorts of insect eggs, so they are useful and highly versatile predators. To encourage them, build log piles, bug hotels and bug boxes for overwintering sites, or make piles of stones. A small nettle patch will attract those ladybird species that are found on nettles, while a selection of early flowers will provide a supply of nectar and pollen.

Beneficial bugs

Most of the bug family (Hemiptera) are plant eaters, but there are some beneficial predatory species, the most notable being the assassin bugs. Most are tropical, but there are seven species in the UK. Most useful is the thread-legged bug (*Empicoris vagabundus*), which is found on trees and shrubs and eats greenfly and other small insects. Flower bugs (*Anthocoris nemorum*) are common on trees and shrubs in gardens. Just a few millimetres in length, they prey on soft-bodied insects such as aphids and thrips. Sometimes, they are considered a pest, as they are found around fruits, especially autumn-fruiting raspberries, and bite people.

Natural predators

Hoverflies

Regular visitors to fields and gardens, hoverflies are seen from late spring to autumn, flitting around flowers as they feed on nectar and pollen. Hoverflies are a fascinating group of insects, mimicking the appearance of bees, wasps and even hornets. It's an example of Batesian mimicry, when an animal looks like a poisonous animal, but is not poisonous itself, using the subterfuge for protection. The black and yellow colours of the hoverfly deter would-be predators, even though they are completely harmless. Hoverflies are true flies and are able to hover and dart sideways. It's this style of flight that sets them apart from the bee or wasp. There are around 280 species of hoverfly in the UK. They are important pollinators but, more importantly, more than half the species have predatory larvae that feed on aphids, thrips, scale insects, mites, etc. The larvae are far less noticeable, resembling small green slugs.

Lacewings

Green lacewings are seen when there are plenty of aphids or whitefly around. They have pale green bodies, large transparent wings and golden eyes. The adults feed on nectar, pollen and honeydew, while the larvae are predatory. Lacewing larvae are small, with a grey-brown body and huge mouth parts, often described as ice tongs, which they use to grab prey such as thrips, aphids, whitefly and spider mites. They are active predators, reportedly eating up to 200 aphids, whitefly or other prey every week. The adults overwinter in sheds, log piles, bug hotels and other sheltered places.

Parasitic predators

I once read an article that described the actions of the parasitic wasps as being far worse than any science-fiction movie! The scene of the creature bursting out of its living host in the film *Alien* may seem far-fetched, but it is happening every day in your garden, albeit on a much smaller scale. There's a vast range of parasitic

Pests and predators

predators, or parasitoids, in the garden. The adult lives a normal, independent life, but their larvae hatch inside another living organism, consuming it and growing. This benefits our gardens if the host is a pest species. Usually, the host dies as or before the parasitoid emerges to live their adult life. Among the most common parasitoids in the garden are parasitic flies and wasps. Some parasitise a range of hosts, while others seek out specific hosts. The larvae feed on the host and then pupate, either inside or out, before emerging as an adult. The adults generally feed on nectar and pollen.

Parasitic flies

These are not the annoying houseflies and bluebottles, but bristly flies that are mostly grey-black in colour, with short antennae and huge eyes, but no biting mouth parts. They lay their eggs on leaves, so the host either eats the eggs or the newly hatched larvae as it feeds. They parasitise hosts such as caterpillars, sawfly larvae and leatherjackets. Once the host dies, the larvae continue to eat the remains before pupating and overwintering. One of the UK's largest flies is *Tachina grossa* which parasitises caterpillars; it is so large that it is often mistaken for a black bumblebee.

Parasitic wasps

Wasps and bees belong to the order Hymenoptera: insects with two pairs of wings and a distinctive narrow waist. There are honeybees, bumblebees, solitary bees and social wasps, but remarkably, of the UK's 7,761 species of Hymenoptera, around 6,500 are parasitoid. Parasitic wasps vary enormously in size, ranging from a millimetre to 30mm (1¼in) long. Many females have an ovipositor, a sting-like attachment at the end of the abdomen, which is used to pierce the body of the host and lay eggs. Their hosts are mostly butterfly and moth caterpillars and pupae, and sawfly larvae.

It's easy to spot the larger parasitic wasps, but the small wasps tend to go unnoticed. Both inhabit our gardens. Like parasitic flies, the parasitic wasps can attack the egg, larval and adult stages of the

Natural predators

host. Once the parasitoid egg hatches, the larvae feed on the host's body tissues, taking the fat stores and non-essential organs first, as they don't want the host to die too soon. Once their larval stage is complete, they pupate. Some will pupate outside the dead host, others inside. Among those most commonly found in gardens is the tiny braconid parasitoid *Cotesia glomerata* that looks like a flying ant and parasitises the caterpillars of the large white butterfly.

Chalcid wasps

Chalcid (pronounced 'kal-sid') wasps are just 3–9mm (⅛–⅓in) in length. You can identify them from the swollen femur on the hind

The zombie ladybird

The common, but rarely noticed, green-eyed wasp (*Dinocampus coccinellae*) parasitises the adult seven-spot ladybird. A fascinating research paper revealed that this wasp injects a virus into the host as well as an egg. The larva feeds on the ladybird but doesn't kill it. A few weeks later, the larva bursts out of the ladybird and spins a cocoon between its legs. The ladybird doesn't react at this point because its brain has been attacked by the virus, causing it to be paralysed and become a 'zombie bodyguard' – an example of a neurological weapon! If you see a ladybird that is alive but not moving, check underneath for a cocoon. It may be a zombie! Even more remarkably, one in four ladybirds will recover from this ordeal. A truly astonishing relationship!

pair of legs and their glossy metallic colours. Because of their size, the number of chalcid wasps is often underestimated, but they are incredibly useful. They lay their eggs in the eggs and larvae of flies, beetles, moths, butterflies, leafhoppers, thrips and scale insects. Occasionally, you may spot one tapping leaf surfaces with its antennae in search of their host's 'scent', but the presence of sick or dead hosts is a sure sign you have a healthy population of chalcids. Many of the smaller parasitic wasps are used as biocontrol in glasshouses (see 'Using parasitic wasps', page 121).

Social wasps

Not the most popular of insects, but we need to encourage this insect because it's such a useful predator! There are eight species of social wasps in the UK, and they are both pollinators and predators. Their life cycle starts in autumn, when a mated queen hibernates in holes while the rest of the colony dies. She becomes active in spring, finds a nest site, and starts building the nest and laying eggs. Four weeks later, the first generation of workers emerges. The workers hunt for insects, especially caterpillars, to feed to the larvae, while the workers themselves feed on the sugar-rich secretions of the larvae. Social wasps are valuable predators in the garden, as they help to control the numbers of insect pests. But, by late summer, there is a decline in the number of wasp larvae and the supply of food for the workers dries up. Then they have to seek out sugar, which results in damaged fruits and conflict with people.

Spiders

Spiders are generalist hunters and have a varied diet. One important fact is that they tend to kill more prey than they eat, so they are incredibly useful in spring when they can limit their prey's early population growth and generally exert a stabilising effect. A garden can have a surprising diversity of spiders, which is important since the failure of one species won't affect pest control if there are others to take its place. You can boost spider diversity by covering the soil with mulch, especially straw, and building log piles and bug hotels.

Centipedes

The fast-moving centipede is found in leaf litter, under pots and logs, and in the compost heap. Centipedes use their jaws to inject a paralysing venom into their prey. It's easy to distinguish between the centipede and the closely related millipede, as centipedes have a single pair of legs per segment while millipedes have two. Even millipedes are not harmful; they mostly eat dead and decaying matter, which is why they are found under the bark of rotting logs and in the compost heap, so they have a useful role to play. It's only occasionally that they will be tempted to feed on seedlings.

Earwigs

This invertebrate has a poor reputation but, on balance, it is broadly beneficial and adds to the predator diversity in the garden. Many gardeners trap earwigs with upturned clay pots on sticks stuffed with straw, especially around dahlias because they can damage the flower buds, but their benefits far outweigh their disadvantages. They are scavengers, predators and pollinators. They feed on dead and decaying matter, for the most part, but they will also prey on aphids, snails and other small pests and help to control codling moth on fruit trees. If earwigs are causing a problem around your young crops and flower beds, use live traps and move them to a log pile.

Amphibians

Toads prey on slugs and snails, grasshoppers, ants, flies and other invertebrate animals. They live away from water as an adult, digging out a shallow burrow in which to shelter and overwinter in deep leaf litter and log piles. You will also find toads in compost heaps, so it's essential to do a toad check before you stick your fork in it. Frogs are similar in many respects to toads but are more agile, and they eat a similar range of prey. They hibernate over winter in pond mud or under piles of logs, stones and leaf litter. Both frogs and toads need water to breed so, if you want to encourage frogs and toads, then building a small pool is the answer.

Reptiles

If you live in a more rural area, it's not uncommon to find grass snakes, slow worms and lizards in the garden. I have found grass snakes under piles of compost in the polytunnel, and allotment holders frequently report seeing them under sheets of metal and finding their eggs in compost heaps. Grass snakes are useful predators, eating insects and small mammals, but they do take frogs and toads, which make up the bulk of their diet. Slow worms emerge from their shelters at dusk or after rain, to prey on insects, slugs, snails, worms and spiders.

To encourage reptiles, you need to provide an open sunny spot where they can bask in the sun. I put down corrugated metal sheets and broken pieces of slate, which grass snakes and slow worms can sun themselves on or shelter under. Lizards enjoy a sunny wall or a pile of rocks in a sunny spot. The heat of a compost heap will attract the reptiles, too. Think about hibernation places as well, such as an undisturbed pile of rocks, log pile or pile of leaves.

Birds

Birds are a sure sign of a heathy garden, but the numbers of many bird species have fallen dramatically in recent years, especially insect eaters, fuelled in part by the collapse of their food chains. Just as having a diverse array of insects in the garden is vital, so it's important to have a wide variety of birds to occupy different niches and eat different foods. Their key role in a healthy garden is to keep a lid on the pests. Insect eaters, such as blue tits, robins, house sparrows, fly catchers, dunnocks and wrens, are particularly important as they will eat aphids, bugs, caterpillars, spiders and woodlice. When feeding nestlings, adults need high-quality food, so they often collect hundreds of caterpillars a day. Blue tits, in particular, tend to time their egg laying to the bud burst of the oak

Natural predators

> ### Attracting more birds
>
> **Cover** Birds need places to shelter from the weather and to nest, so trees and shrubs are ideal. If you have the space, establish a shrubby area and grow climbers on a wall or fence. On farmland, you get more birds around the field margins, especially if they are woody, so establish a diverse boundary hedgerow to draw in the birds.
>
> **Layered habitats** Have a mix of tall, medium and low-growing plants to provide shelter and cover.
>
> **Nest boxes** Providing nest boxes will bring in insect eaters like robins and blue tits.
>
> **Water** This is essential all year round, especially in winter when water sources may freeze.
>
> **Food** In winter, put out an array of different feeders and foods that will appeal to as wide a range of birds as possible. It's best not to feed birds in summer, as you want them working for you on pest-control duty. Make sure your trees and shrubs are species that provide birds with berries and fruits.

trees, as they harbour a huge winter moth caterpillar population. Research shows that the winter moth caterpillar is the ideal food for these birds, but most gardens have a low capacity to supply the nestlings with enough food. Alongside caterpillars, blue tits also feed on greenfly, but these are not as good as moth caterpillars.

Pests and predators

Urban blue tits tend to feed more on ladybirds and spiders and, as a result, they have less breeding success than rural ones. Birds need a good supply of calcium for eggs, so small birds like robins and blue tits, which have limited storage of calcium, can also be found foraging on the ground for woodlice, millipedes and spiders, which contain calcium.

In the UK each year, in winter, the Royal Society for the Protection of Birds (RSPB) holds the Big Garden Birdwatch, asking people to count the birds they see in their garden. In 2023, more than half a million people counted almost 8 million birds with the top 10 being (from the top) house sparrow, blue tit, starling, wood pigeon, blackbird, robin, goldfinch, great tit, magpie, and long-tailed tit.

But there are some alarming trends for some of the birds in the top 10. Since the start of the count back in 1979, house sparrow numbers have fallen by more than half, and starlings are down by a worrying 80 per cent, blackbirds by 46 per cent and robins by 32 per cent. The reasons include fewer green spaces, less available food, pollution and the effects of climate change.

Hedgehogs

Hedgehogs are great for controlling pests, as they feed on slugs, beetles and larvae of all sorts. Their numbers have plummeted in recent decades, but they are beginning to return to urban gardens, and you can encourage them by providing log and leaf litter piles for hibernation and shallow water sources to drink from. They travel long distances each night in search of food, so providing hedgehog holes (13cm/5in square) in your boundary walls and fences, allowing them to move from garden to garden, is essential.

Bats

I love watching bats at dusk as they flit around the house preying on moths, midges and other insects that are attracted to our lights. To get bats, you need lots of insects, so build a pond and establish a sequence of flowers (see 'Planting for natural predators', page 82). They are attracted to pale flowers that release scent at night, so you

Natural predators

can encourage them by planting ivy, summer jasmine and honeysuckle on walls and fences. Bats use trees and hedgerows to navigate and for shelter, while they use rot holes in mature trees to roost. They prefer large, mature trees, such as ash, oak and beech, but any tree is better than none. If you lack bat-roosting sites, you can put up artificial bat roosts; either make your own or buy one ready-made. Like bird boxes, each species of bat needs a different shape of box.

WAYS TO ENCOURAGE PREDATORS

Here are five ways to encourage natural predators to stay in your garden.

Beetle banks and buckets

You need to provide ground beetles with shelter, so they don't wander from your garden. Farmers encourage beetles by building beetle banks, which are raised banks planted with grass that stretch across fields, to provide shelter. You can create something similar in your garden by building a small bank running along the side of the vegetable patch or even a raised circular bed, around 1m (3ft) across, planted with a mix of ornamental grasses. These raised grassy areas attract other predators, too, including spiders, ladybirds and solitary bees.

If you don't have space for a beetle bank, you could opt for a beetle bucket. Take an old bucket and cut small circular holes in the sides and bottom so beetles can get in. Bury the bucket so the top is level with the ground and fill with logs, twigs and fallen leaves. Then top it with some larger logs, which will protect the habitat below.

Bug hotels

There are many kinds of 'bug hotels' that provide shelter and overwintering spots for all sorts of invertebrates and vertebrates. Most are made using pallets that are filled with a mix of natural materials, including bundles of hollow-stemmed grasses, pinecones, roofing tiles, logs with holes, crocks, moss, wool, and so on. I usually make mine from old pallets, but these do rot after a few years, so why not make yours from more unusual and long-lasting materials, such as gabion baskets which are made of galvanised wire mesh. Gabion baskets are usually filled with rocks to create a barrier such as a wall, but three of the smaller baskets are perfect for filling with nesting materials and you can then stack them on top of each other to create an eye-catching garden feature.

Mulching

A mulch of leaf litter, compost and straw provides the perfect overwintering site for spiders, ground beetles and ladybirds.

Wood mould boxes

Look inside a cavity in an old tree and you are likely to find a pile of decomposing leaves, twigs and bark. Not surprisingly, this is home to many invertebrates, too. You may not have an old tree in your garden, but you could recreate this vital habitat by building a wood mould box to mimic a tree cavity. Make the box from planks of wood – an ideal size is 70cm (28in) tall by 30 × 30cm (12 × 12in) to create a cavity of 60 litres (16 gallons). Line the bottom of the box with some clay to trap moisture. This is best done by dropping some wet clods of clay into the cavity and using a piece of wood to push it down into the corners. Then fill three-quarters of the cavity with oak leaves, hay, bits of bark, feathers, old birds' nests, moss, lichen, etc., and add 5 litres (1¼ gallons) of water to boost the moisture level. Secure the roof in place. Drill some small holes in the sides to let in rain and attach the box to a garden tree, on the north or east side, so it's not in full sun.

Wood piles

Dead wood is important for wildlife. Don't cut up waste branches; instead, simply pile up a mix of large and small branches in a shady place. This shouldn't be dense shade, as dappled shade is perfect. Adding some leaf litter can attract an even greater range of animals.

Natural predators

Orchards

There are many pests in orchards, so you need a healthy population of natural predators to keep them in check. Your main allies are:

Ladybirds Attract these by planting a range of nectar-rich plants in and around the orchard and building a bug hotel for overwintering sites.

Lacewings These eat aphids and are attracted by nectar-rich plants.

Predatory bugs (capsids, nabids and flower bugs) These prey on a wide range of pests, such as aphids and codling and tortrix moths. Encourage the bugs by providing overwintering spots, including artificial bottle refuges. These can be made by taking a 1-litre (34-fl oz) plastic bottle, cutting off the bottom and filling it with a roll of corrugated card. Then attach the bottle horizontally to a tree with tape. Ideally, provide one bottle per tree.

Parasitic wasps Laying eggs on aphids, caterpillars and sawfly larvae, these wasps can be attracted by nectar-rich plants.

Ground beetles Tackling sawfly larvae and caterpillars on the ground, these beetles need shelter, so put down some logs.

Spiders These will eat a wide range of orchard pests. A single apple tree may be home to as many as 10 different species of spider.

Pests and predators

Hoverflies and larvae These control aphids, including woolly aphids. Allow the grass to grow long and encourage plants such as yarrow, cow parsley and hogweed.

Earwigs These predate woolly apple aphids, codling moth caterpillars and scale insects. A refuge can be formed from a plastic bottle.

Birds Encourage birds such as blue tits, great tits, sparrows and robins, which help to control the numbers of aphids and caterpillars. Put up nest boxes to boost their numbers.

Bats These will take codling moth adults, so encourage them by putting up Schwegler woodcrete bat boxes on tree trunks to provide a roosting place.

PART 3 Plant diversity

CHAPTER 5

Getting the planting right

In nature, ecosystems that are more diverse show greater stability and resistance to withstand disturbance and increased resilience to recover from changes, coping with all that life throws at them – including extreme weather and outbreaks of pests and diseases. So, to create a more resilient garden, you need diversity. This will also put predators at an advantage, by providing them with plenty of food and places to shelter.

Looking back at gardening history, the typical 14th-century cottage garden, with its mosaic planting of ornamental and edible plants, made it more difficult for pests to find their host plant among a swathe of scented plants, while the sheer diversity of plants, of course, attracted an equally diverse selection of predators and pollinators. Such a garden style didn't allow one species to get out of control and its numbers to increase.

However, by the time we reached the Victorian period, fruit trees and vegetables were not considered suitable for ornamental beds and they were moved to the kitchen garden, where crops were grown in long rows for ease of management and the soil was hoed so it was weed- free and bare. It is hardly surprising that they suffered from pests and disease and, in true Victorian style, the gardeners introduced all manner of chemical controls.

Plant diversity

For the modern gardener, it's far better to have a potager or ornamental kitchen garden where you can grow your crops among the flowers. Not only does this look wonderful, but the garden will also suffer less from pests and disease.

PLANTING FOR NATURAL PREDATORS

You want to keep natural predators in your garden for as long as possible, so you need to bribe them to visit and then stay. This is achieved by ensuring you have a sequence of flowers supplying pollen and nectar for as much of the year as possible.

Choosing the right flower

When choosing plants to attract predatory insects, you need to bear in mind that many of these insects – especially hoverflies – have specialist mouth parts that cannot push deep into a flower to reach the nectaries, unlike bees and butterflies. They need small flowers with exposed nectaries rather than large, tubular flowers. Often, it's best to go for a mix of flower types when choosing plants for the herbaceous bed, so that they can service a wide range of insects, from the pollinators to the predators.

There's a huge range of flowers to consider, and many can be established from seed. Among those I consider to be the most useful for attracting predators are *Angelica* spp., borage (*Borago officinalis*), buckwheat (*Fagopyrum esculentum*), coriander (*Coriandrum sativum*), fennel (*Foeniculum vulgare*), *Phacelia tanacetifolia* and sweet alyssum (*Lobularia maritima*). You can let brassicas and parsnips overwinter and flower. I always have a diversity of flowers so that, should one type fail, there are plenty of others to compensate.

Hoverflies are essential predators, and their numbers are boosted by planting nectar-rich flowers, but the choice of flower is critical. Predatory hoverflies have shorter mouth parts than the non-predatory species and research has found that they need a flower depth of just 1.6mm, which is less than the length of their

Getting the planting right

Table 3. Plants to attract specific predatory insects

PREDATOR	PLANTS TO GROW
Hoverflies, predatory	Angelica, buckwheat, chamomile, coriander, dill, elder, fennel, hogweed, knapweed (Centaurea), parsley, phacelia, yarrow
Lacewing	Angelica, caraway, coriander, cosmos, cow parsley, dandelion, dill, Dyer's chamomile (Anthemis tinctoria), fennel, parsley, sweet cicely, yarrow
Ladybird	Ajuga, angelica, calendula, caraway, chives, coriander, cosmos, dill, fennel, feverfew, French marigolds, nettles, sage, sweet alyssum, yarrow
Tachinid flies	These have sponge-like, not tube-like, mouth parts, so need shallow flowers such as angelica, aster, chamomile, coreopsis, coriander, dill, shasta daisy and yarrow.
Wasps, parasitic	Anise hyssop (Agastache foeniculum), brassicas, coriander, dill, fennel, parsnip, phacelia, tansy
Wasps, social	Figwort and nectar-rich flowers

mouth parts. This means that you need to grow more open flowers where pollen and nectar is easily accessed, such as angelica, dill, fennel and parsley (see 'Table 3. Plants to attract specific predatory insects'). Hoverflies can reach nectar in other flowers if there is enough room for them to push their head inside to reach the nectaries. Many gardening books recommend cosmos (*Cosmos* spp.), poached egg plant (*Limnanthes douglasii*), pot marigold (*Calendula officinalis*) and sunflower (*Helianthus* spp.) for hoverflies, but their nectar-producing florets are tightly packed together and inaccessible for the predatory hoverflies. A weed you may not

Plant diversity

welcome, but which is great for attracting hoverflies, is hogweed (*Heracleum sphondylium*), although some people can get a skin rash from touching it.

Another essential insect is the parasitic wasp. North American studies of one species that parasitises whitefly on beans discovered that the tiny wasp visited almost 100 nectar-producing plants close to the crop, showing that you really can't have enough diversity of flowers near your vegetables.

MY TOP 10 EASY-TO-GROW FLOWERS

There are some flowers that I'm never without and, fortunately, they are easy to grow from seed. My top 10 here will provide you with a steady supply of flowers from early spring to autumn. Banks of these flowers around your growing area will make sure that predatory insects don't have to travel far to find food.

Phacelia (*Phacelia tanacetifolia*)

Predatory insects just love this plant. It's brilliant for bees, but studies have shown that phacelia is also the best flower for attracting parasitic hoverflies, ensuring they spend longer on the plot, laying eggs and hunting. Studies in Europe show that corridors of phacelia in sugar beet fields lead to fewer bean aphids due to more parasitic hoverflies, while growing buckwheat and phacelia in Swiss cabbage fields increased populations of parasitic wasps that attacked cabbage aphids. Phacelia seed is easy to sow; simply scatter seeds when you have a patch of bare ground in late summer and autumn and it will provide you with early spring flowers, then sow again in spring for later in the year. And if you allow it to self-seed, you will have a continuous supply of flowers across the plot. Learn to recognise the frilly leaves and you can simply hoe the seedlings you don't want.

Borage (*Borago officinalis*)

It's claimed that the borage plant can produce nectar more quickly than any other plant. I don't know if that's true, but I can safely say

Getting the planting right

my borage in early spring buzzes with bees of all sorts, as well as ladybirds and other predators. Once sown, you don't ever need to sow again unless you are a very tidy gardener. Mine set seed all over the place, so it's just a question of which ones to leave. They grow into large, straggly plants, so this determines where I let them grow. It's also a useful companion plant (see 'Companion planting', page 88), producing flowers from early spring to late autumn.

Parsnip (*Pastinaca sativa*)
In the late spring, the open umbels of yellow flowers really bring in the insects.

Buckwheat (*Fagopyrum esculentum*)
This plant flowers for the longest period, up to three months, which can take you through to late summer when many other flowers have given up. Buckwheat is drought tolerant, so it also helps in a dry summer when the lack of water can wipe out your annual sowings. Sow in late spring to early summer for flowers from early summer to early autumn.

Sweet alyssum (*Lobularia maritima*)
This low-growing annual creates a carpet of white-scented flowers. It's easy to establish; simply scatter the seeds on a patch of bare soil. It flowers from late spring through to late summer. Not only does sweet alyssum bring in predators, but the carpet of plants protects the soil in summer with a living mulch and reduces water loss.

Angelica (*Angelica* spp.)
This biennial produces the most amazing statuesque flower stalks in spring which attract bees, hoverflies, ladybirds and parasitic wasps. It likes a shady spot with mulched moist soil. Seeds sown in autumn or late spring will grow on through the first year and flower the following spring.

Plant diversity

Sunflower (*Helianthus* spp.)
Who doesn't love seeing the huge yellow heads of the sunflower? They attract a steady stream of insect visitors, not just bees and bumblebees, but also lacewings and ladybirds. But sunflowers aren't great for hoverflies. You can grow many different varieties: some dwarf, others multistemmed, as well as the typical giant sunflower. Sow in spring for flowers from mid-summer to early autumn. There are also perennial members of the sunflower family to consider, including Jerusalem artichoke (*Helianthus perennis*), which flowers late in the season and has the added bonus of edible tubers, and the Maxmillian sunflower (*Helianthus maximiliani*), a tall, drought-tolerant perennial with loads of bright yellow flowers in late summer into autumn that attract both butterflies and birds.

Fennel (*Foeniculum vulgare*)
This can be grown as an annual or perennial. Just like parsnips, its umbels of yellow flowers are great for pollinators in late spring and early summer. It's easiest to let some of your fennel crop go to seed, so you are never without it, or you can buy the more ornamental perennial purple fennel.

Marigolds (*Tagetes* and *Calendula* spp.)
There are lots of different marigolds for the garden, including African, French and pot marigolds. All are easy to sow, most self-seed, and they attract bees and hoverflies. Pot marigolds (*Calendula officinalis*) flower from mid-spring through to the first frosts, while French and African marigolds (*Tagetes patula* and *T. erecta*, respectively) flower in summer through to autumn. My polytunnel French marigolds often continue flowering through to early winter.

Dandelion (*Taraxacum officinale*)
Considered by many gardeners to be a weed, the dandelion is actually a useful plant as it flowers early, providing that essential nectar supply, and the leaves are edible and full of nutrients. Try

Getting the planting right

this idea around your raspberries. Establish dandelions between rows of summer raspberries to attract pollinators. You get a sequence of flowers: dandelions first, then raspberries, followed by the other berries, such as blackberries. Mow the dandelions once they have flowered, so there is no seed to spread. Dandelions are deep-rooted, so they bring up minerals from the subsoil and easily out-compete the other weeds. It's so much easier to let the dandelions control the other weeds than to dig them out of the ground.

WEEDS

Attitudes are changing towards weeds. The conventional definition of a weed is a wild plant growing where it's not wanted. But we are discovering that some so-called weeds can be very useful. If you can live with a few weeds, their presence can lead to more pest control. For example, research shows that outbreaks of some pests on arable farms are less likely in a weed-diverse system than one that lacks weeds. One idea is that the weeds break up the outline of the crop, making it more difficult for the pests to pick out their target. This is a balancing act, of course, for weeds also take up water and nutrients and shade your vegetables. Personally, I feel complete weed control is really undesirable. Rather, you should be looking to manage the weeds in your growing space, as opposed to removing them completely. There are places where you can allow weeds to grow, such as at the edge of your vegetable patch and along boundaries, where their flowers can attract pollinators and predators, and you can crowd the weeds out where you don't want them by planting your crops closer together. Also, a carpet of low-growing weeds, such as chickweed (*Stellaria media*) or ground ivy (*Glechoma hederacea*), can form a perfect living mulch (see 'A living mulch', page 37). Just remember to stop the weeds setting seed or mow or chop them off once they have achieved your aim; otherwise, at the end of the season, they will sneak in another generation of flowers and seeds!

COMPANION PLANTING

This is where you grow one type of plant, secondary to the crop, to repel, disrupt, trap or lure pests and pathogens as well as to attract or provide food for predators. It's not a new idea as records show that the ancient Greeks and Romans knew about companion planting, as did the indigenous peoples of North America.

Nowadays, the term companion plant tends to be used as an umbrella term for a plant combination that benefits the crop. It's most commonly used in gardens and market gardens, but is increasingly seen in field-scale operations, too. A lot has been written on this topic, but you have to be careful, as I have found that much is based on anecdotal evidence and many companion pairings don't really work. For example, planting basil near tomatoes to deter whitefly, and growing onions near carrots so that the smell of the onion deters the carrot root fly, fails to work for me. I have sought out planting combinations that research has proved to be effective, but the real benefit of companion planting is the boost in biodiversity.

REPELLING PESTS

You can use companion plants to keep the pest or pathogen away from your crop. This can be achieved by planting an unpalatable plant around the crop to make it more difficult for the pest to move between the crop plants, so they don't eat as much. Or you can confuse the pest by growing a companion plant that releases an odour that masks the presence of the crop. One valuable companion plant is the French marigold (*Tagetes patula*) which has long been planted near tomatoes to deter whitefly through its smell. There is now evidence that the roots of this marigold produce a chemical, alpha-terthienyl, that acts against nematodes, insects and even viruses, while the presence of this chemical inhibits the hatching of nematode eggs, although it's not clear how it works. So, an additional benefit of planting French

Getting the planting right

marigolds around tomatoes in greenhouses is to reduce the number of root-knot nematodes that form large galls on tomato roots, reducing water and nutrient uptake and leading to wilting and stunting. Researchers are still trying to work out what's going on, but it's possible that the marigold roots are acting as a trap for the nematodes.

I grow a lot of French marigolds in my polytunnel. I used to collect the seed and clear them away, but now I leave them where they are to rot down and allow their decomposing remains to release chemicals that not only inhibit the germination of weed seeds, but also deter harmful soil nematodes. There is one pairing I have not yet tried, which is planting marigolds around potatoes to deter slug damage late in the season. Perhaps next year.

TRAP CROPS

In trap cropping, the companion plant is positioned near the crop to lure the pest away. Many gardeners already do this when they plant nasturtiums to lure white butterflies away from brassica plants or to attract blackfly. French marigolds are also effective traps, as they attract thrips and red spider mite. Farmers usually grow the trap crop as a border around the perimeter of the field because many pests overwinter in hedgerows and move into a crop from the edge. A trap crop can also be sown between rows of crops. But remember that the trap plants have to be pulled up to remove the pests and, if necessary, resown to maintain the attraction. One widely grown trap crop is mustard, which is particularly attractive to flea beetles. Flea beetles can be problematic in the garden, affecting brassicas at the young stage. The leaves of my rocket and broccoli plants are usually peppered with tiny round holes. To gain protection, sow some strips of green manure mustard seed around the edge of your vegetable area or allotment. Also, you can intercrop your brassicas with radish as a sacrificial crop. Radish is also a brassica, so it will attract the flea beetles away from your main crop.

Plant diversity

In North America, cucumbers often suffer infestations of cucumber beetle. Trials using a trap crop of pumpkin and squash found the cucumber yield increased by 18 per cent and insecticide use fell by 96 per cent. An extra benefit was a small harvest of pumpkin and squash that had survived the beetle and could be sold as a secondary crop. The squash variety 'Blue Hubbard' was found to be the best – the beetles didn't appear to like the fruit, so they didn't bother to move through the border of squash to find the cucumber plants beyond.

Push-pull pest management

This takes the trap crop idea one stage further. In a push-pull system, the main crop is interplanted with a crop that pushes the pest away, while the pull crop is planted on the edges of the crop to draw in or trap the pests. For example, carrots are susceptible to aphids, so you might grow chives around them in order to push aphids away and, at the same time, plant nasturtiums to pull the aphids away. If you have problems with flea beetles on your brassicas, interplant with onion, mint or garlic to push the flea beetles away and plant green manure mustard around the edge of your growing space to pull the flea beetles away.

BARRIER CROPS

Simple ideas such as growing tall crops as a barrier or in strips between crops can be very effective in reducing the number of airborne pests like aphids reaching susceptible crops. For example, growing a border of sunflowers, Jerusalem artichokes or quinoa, which are not susceptible to aphids, can reduce the number of aphids reaching a crop that is susceptible.

GROWING DIVERSITY

To an insect pest, a fertilised, weed-free monoculture is pure heaven, while the sheer number of plants growing close to each

Getting the planting right

Plant companions

Researchers have found the following plants to be valuable companions:

Chervil (*Anthriscus cerefolium*) around both vegetables and ornamental plants to attract slugs and snails.
Dill (*Anethum graveolens*) as a trap plant around tomatoes for tomato hornworms (North America).
French and African marigolds (*Tagetes patula* and *T. erecta*) between potatoes, brassicas, legumes and squash to reduce thrips and nematodes.
Horseradish (*Armoracia rusticana*) or tansy (*Tanacetum vulgare*) between rows of potato to attract Colorado potato beetles (North America and parts of Europe).
Legumes, such as beans, around sweetcorn to reduce leafhoppers, leaf beetles and fall armyworms (North America).
Medick (*Medicago littoralis*) between rows of carrot to confuse carrot root flies.
Mustard and radish (*Rhaphanus sativus*) to trap flea beetles in brassica crops.
Nasturtiums (*Tropaeolum majus*) to attract aphids, blackflies, whiteflies, flea beetles, cucumber beetles, squash vine borers and white butterflies.

Plant diversity

> **Nettles** (*Urtica dioica*) to attract aphids.
> **Pot marigolds** (*Calendula officinalis*) around garlic to reduce thrips.
> **Tansy** (*Tanacetum vulgare*) around alliums to reduce incidence of rust.
> **Tomatoes** intercropped with brassicas against diamondback moths or cabbage moths (*Plutella xylostella*).

other over a large area makes it easy for disease to spread. Many pests and diseases have adapted to the way we grow our crops in simple cropping systems, while the natural predators have not fared so well because the system works against them. Tilling, weeding, spraying, harvesting and other farming activities damage habitats for natural predators. Although this describes the scenario on intensively farmed land, the same can be said of many a garden, where the gardener digs the soil, weeds, sprays the pests, harvests crops and leaves the soil bare until the next season. Not surprisingly, we see the most pests and disease in the vegetable plot and orchard.

I like to see diversity in my vegetable plots, so I grow a mix of annual vegetables and flowers alongside perennial vegetables and cover crops, all designed to bring in many different types of beneficials, both pollinators and predators. All the beds are managed using no-dig principles (see 'No-dig gardening', page 22), with an annual layer of compost, woodchip or mulch in winter. There is also diversity in the type of crops: different crop families, annuals and perennials, tall and short, deep-rooting and shallow-rooting, plus diversity within the crops by way of different varieties.

Getting the planting right

ROTATION, POLYCULTURE AND CONTINUOUS CROPPING

Like many gardeners, I read about crop rotations and how this was essential for vegetable growing. But the more I read about the ways in which people in other parts of the world grow vegetables and what some of the more innovative farmers and growers are doing, the less convinced I am about the need to rotate and I have started to move away from a rigid crop rotation to something more fluid.

I remember a phrase used by soil scientist Elaine Ingham at one of her workshops: 'Nature doesn't rotate'. She used the example of old-growth forests in North America that have been growing on the same soil for thousands of years without any lack of nutrients because the soils were rich in microbes and the whole system was fungal-based. She was clear that it was possible to grow the same crop in the same soil each year if conditions for that crop were ideal, quoting an example of corn being grown in the same field for 35 years without loss of yield. So, when it comes to growing vegetables, you have three options: crop rotation, polyculture or continuous cropping.

1. Crop rotations

Traditionally, gardeners have grown their crops in a sequence called a rotation. A crop rotation is said to keep the soil healthy, meet the nutritional requirements of the different crops, reduce weeds and help to reduce the buildup of pests and disease. It also helps to increase soil organic matter and protect the soil while giving good yields.

Reams have been written about how crop rotations balance out the nutritional requirements of the crops, with one crop taking up more of one type of nutrient than another, with deep-rooted crops bringing up nutrients that can then be used by shallow-rooted crops and legumes fixing nitrogen for use by the following crop. But, as one organic grower pointed out to me, legumes may fix nitrogen, but they use it to fuel their own growth and produce

Plant diversity

seeds which you then harvest, so not much will go back into the soil. If you want to boost nitrogen, you should grow a cover crop of clover or undercrop with clover (see 'A living mulch', page 37). And if you add a layer of compost each year, there will be plenty of nutrients.

It's clear that crop rotation helps to disrupt pest and disease cycles, and it's particularly useful when controlling pests and pathogens that overwinter in the soil or crop residue. By having a different crop in the soil each year, you break the cycle. For example, if you have had a carrot crop that was infested with carrot root flies, there will be pupae overwintering in the soil. If you sow carrots in the same place again the following spring, the adult flies will emerge around your new carrot crop. However, if you rotate, they will emerge to find the crop is no longer present. But a rotation doesn't have to be once a year. Many gardeners get more than one harvest a season, so a bed of early potatoes may be followed by leeks, or early salad crops followed by late brassicas.

The two main rotations used by gardeners are the three-crop and four-crop rotations. The three-crop rotation has a lot of emphasis on potatoes and brassicas, so is favoured by many commercial growers: Year 1 (potatoes), Year 2 (legumes, onions and roots) and Year 3 (brassicas). A typical four-crop rotation, with more space for legumes and roots, consists of Year 1 (potatoes or squash), Year 2 (onions or legumes), Year 3 (brassicas) and Year 4 (roots), but there are plenty of variants to suit your needs and space. Many commercial growers have developed their own rotations based on years of testing and observation. Eliot Coleman, an organic vegetable grower based in the US, has rotations running over eight years:

Year 1 Peas
Year 2 Brassicas undersown with clover
Year 3 Sweetcorn
Year 4 Potatoes
Year 5 Squash

Getting the planting right

Year 6   Roots
Year 7   Beans
Year 8   Tomatoes

2. Polyculture

As a novice gardener, all the books I read described growing vegetables in a rotation with neat, weed-free rows and I saw this first-hand in many of the gardens that I visited. But, over the years, I have moved to more of a polyculture approach. Most of my crops are intercropped; that is, I grow small blocks or rows of crops between other crops. This mixing up of the crops makes it more difficult for pests to locate their host plants and disease to spread. For example, a bed may be mostly climbing peas and beans, but I plant courgettes underneath and squeeze in a few parsnips, celery leaf and tomatoes at the end of the bed. Then in another bed I may have alternating rows of spinach, celeriac, parsnips and beetroots with plenty of chives, cosmos and French marigold. I still rotate some crops, so it's a hybrid rotation-polyculture approach.

I tend to grow my brassicas in a single bed, as I cultivate them under butterfly nets to protect against white butterflies and pigeons. So, while I have just one crop type in the bed, I'll plant a mix of brassicas, rather than all the same kind. I make good use of vertical spaces by popping a plant support in the middle of a bed to grow climbers, such as the attractive Malabar spinach or sweet peas. There are arches over paths to link beds, over which I grow climbing French beans, tromboncino squash, sweet peas (*Lathyrus odoratus*) and hopniss (*Apios americana*). My beans and peas are not moved every year, but I move them around within the bed and mix up the varieties, moving them to a new bed every four or five years. It's different with the riskier crops, such as carrots, potatoes and onions. I avoid planting them in the same bed for two years running, so they will be rotated to another bed.

It's easy to incorporate polyculture into your rotation. You can achieve this simply by intercropping your brassicas with a

Plant diversity

fast-growing groundcover crop, such as lettuce or radish. By the time the brassicas have grown, the radish and lettuce will have been harvested. You can do the same with parsnip, which is really slow to get going. Intercropping doesn't usually affect your yields. In fact, the benefits of the groundcover and the different demands on the soil mean that the overall yield is likely to be larger.

Mixing varieties

If you want to keep to a crop rotation but need more diversity, another option is to grow several varieties rather than one, providing within-species diversity. I am sure many keen vegetable growers do this already, as it's not difficult to choose a selection of lettuce, beetroot and carrot varieties, for example, which will have different levels of resistance to pests and disease. If the weather is a bit extreme, you will usually find that some varieties do better than others.

Extreme polyculture

Intercropping is one approach to polyculture and it's easy to implement. But the purists will argue it needs to be more random. So, one year, I took some of my old seed packs – a mix of salads, herbs, spring onions, brassicas and beans – and scattered the seed over a prepared seedbed. Then I waited to see what would happen. The salad crops appeared first, followed by the brassicas and then the beans. I thinned out where necessary and netted the lot, as the brassicas were attracting white butterflies. For the space, I got a lot of crop to harvest, but the random nature of the planting meant that some crops thrived but others didn't – for example, the taller brassicas shaded out some of the smaller plants.

Planting guilds

Another example of polyculture is the planting guild. This is an extension of the companion plant idea, as it's a group of plants grown together, each bringing different benefits to the relationship.

Getting the planting right

It mimics what takes place naturally in ecosystems, such as a forest where there are different layers, with each plant having a specific role and not competing with but supporting others.

A forest garden is the classic example of planting guilds. You start with the focal plant – the tree, which forms the canopy and provides a crop, such as fruit – with secondary plants, like soft fruit bushes and hazelnuts, forming an understory. Next is a nitrogen fixer, which could be shrubs, such as liquorice, lupins or groundcovering clovers. Then, add to this mix a deep-rooting plant, such as comfrey or dandelion, to mine minerals from the soil and you have the bones of a forest garden.

On a much smaller scale, you can establish a planting guild in your vegetable beds. The best known is the 'three sisters' – corn, squash and climbing beans – three nutritious crops that can be stored. The corn is tall and provides support to the climbing beans which fix nitrogen. The squash clambers over the ground, its large leaves keeping the soil cool and moist. You can build on this three-way partnership by adding more plants, such as tall sunflowers to attract pollinators or comfrey to supply minerals – before you know it, you have a polyculture bed. I like to add nasturtiums and pot marigolds, too.

3. Continuous cropping

Natural agriculture takes a completely different approach. It's a system of growing based on the teachings of Mokichi Okada in Japan in the 1930s and is practised in the Far East and increasingly around the world. The idea is to support nature by using no fertilisers, manures, chemicals or pesticides, with the crops grown continuously in the same bed for many years. Followers of natural agriculture believe that if the soil is healthy, it regenerates itself. The soils are never fertilised, even with organic fertilisers. Mulching is allowed, but only with materials from around the vegetable beds, such as dead grass, leaves and crop remains. This material is chopped up and worked into the soil, its role being to break up the soil, warm it and trap moisture, but not feed it. Seed

Plant diversity

saving is important, too, so the crops become adapted to the soil and local conditions.

I have been lucky enough to visit the organic Shumei Natural Agriculture Farm, near Bath, to see this holistic system of growing in action. Their potato harvest is a great illustration of the success of these methods. Over a period of 10 years, the yield of potatoes has increased substantially, despite being grown in the same bed each year. But you can't switch to this style of growing and expect immediate results. It takes time for the soil to adapt to the lack of fertiliser inputs and each year that you save seed, you move away from seed that was bred for growing on a fertilised, intensive system towards one that is adapted to the soil and local conditions (see 'Seed saving – breeding your own resistance', page 110).

If you fancy trying this style of growing, ensure you prepare the soil carefully with some bokashi compost (see 'Fermenting compost', page 30), which is mixed into the soil to ensure it is vitalised and full of beneficial microbes.

I have also seen Charles Dowding's continuous cropping experiments at his no-dig garden in Somerset, just a few miles from me. These have been running for more than 10 years now and he has found that there is no loss in productivity nor any upsurge in disease or pests. Each year, he adds a topdressing of compost (just a centimetre or two) and there has been no loss of productivity or health over this time.

BRASSICAS – A SPECIAL CASE?

I was impressed by the brassica beds at the Shumei Natural Agriculture Farm and have set up my own trial of continuously grown brassicas, now in its fifth year. Unlike most other crops, brassicas are non-mycorrhizal (they don't have a mycorrhiza associated with their roots), so they don't benefit from a fresh topping of fungi-rich compost. Instead, I have been mulching the brassica

Getting the planting right

Timing to avoid pests

If you have a problem with a pest or disease, you can try to time your sowings or planting out to avoid the peak time of the pest or disease. What you want to try to avoid is having vulnerable young crops during the peak time for the pest. For example, spring-sown broad beans tend to have more lush growth and are more susceptible to blackfly compared with those sown in the autumn. It doesn't solve the problem 100 per cent of the time, but it does reduce the risk. Direct sowing needs to be done when conditions are right, in order to ensure the seedlings get a good start, so often it's better to sow in modules and transplant when the plants are well grown. Think about planting out a week or two early, so the crop has a chance to mature before the peak time of disease or pests. For example, a fast-growing early pea can be harvested before the pea moths get going. For all of these options, recordkeeping is important, as you can look up dates from previous years.

beds with grass clippings, hay and poultry manure during the winter, all rich in nitrogen to encourage bacterial growth as well as helping with water retention. The thick mulch can encourage slugs, so I spray with nematodes (see 'Table 4. Beneficial nematodes', page 122). One reason that gardeners are advised to rotate brassicas is to avoid soilborne disease, especially club root, but so far, I haven't had any disease. Only time will tell.

A PERENNIAL PLANTING

For decades, the only perennial vegetables in the plot were asparagus, globe artichokes and rhubarb, but that's all changing as there is a wide range of perennials now available to the gardener. I grow both Chinese artichoke (*Stachys affinis*) and globe artichoke (*Cynara cardunculus*), Jerusalem artichoke (*Helianthus tuberosus*), Good King Henry (*Chenopodium bonus-henricus*), lovage (*Levisticum officinale*), perennial kales (*Brassica oleracea* Ramosa Group), scorzonera or black salsify (*Scorzonera hispanica*), sea beet (*Beta vulgaris* subsp. *maritima*), sea kale (*Crambe maritima*) and skirret (*Sium sisarum*). I also grow a wide range of perennial onions that appear every year and provide me with a constant supply over the winter months, including Babington leek (*Allium ampeloprasum* var. *babingtonii*), potato onion (*Allium cepa* var. *aggregatum*), walking onion (*Allium cepa* Proliferum Group) and Welsh onion (*Allium fisulosum*). The lack of soil disturbance means that leaf litter builds up on the surface; there's a healthy worm population; and plants can grow large root systems supported by mycorrhizae, all of which creates a healthy soil food web. Combine that with the fact that they are low-maintenance plants, and perennials become an essential element of any modern vegetable garden.

AGROFORESTRY

Years ago, I was taught agroforestry at university by a lecturer who had spent many years in Sri Lanka studying the tea plant (*Camellia sinensis*). He explained that they grew trees among the tea bushes to create shade, shielding the tea plants from the extreme sun and rain, and there was no loss in productivity. Over the last 20 years, agroforestry has become more common in the UK, with farmers dividing up their arable fields with rows of trees – usually willow, hazel and fruit trees – to create alleys that are wide enough to continue to use large machinery but provide the farmer with other benefits. One of the best examples is Wakelyns, an organic

Getting the planting right

agroforestry farming hub in Suffolk, where the 23 hectares (57 acres) of farmland is divided into cropping alleys by rows of willow and hazel (which are used for fuel and weaving) and by fruit trees, including apples, pears and even apricots and peaches. The crops are just as diverse and include wheat, lentils, chia, peas and camelina. I have seen successful agroforestry schemes on much smaller market gardens, too. At Tolhurst Organics in Berkshire, and Abbey Home Farm near Cirencester, rows of fruit trees divide the vegetable beds. The trees are underplanted with annual flowers for pollinators and bulbs.

I have incorporated these ideas into my own growing spaces. A row of cordon apples (fruit trees trained as a single stem) along the edge of the vegetable beds provides a crop, a habitat for beneficial predators, flowers for pollinators and shelter from wind. There's a row of crab apples and hazel along the boundary of the pig pens to cast shade in summer and, along another boundary, I have an edible hedge with a mix of autumn olive (*Elaeagnus umbellata*), blackthorn (*Prunus spinosa*), dog rose (*Rosa canina*), hawthorn (*Crataegus monogyna*) and hazel (*Corylus avellana*). This hedge took barely three years to establish and thicken up, yet it not only provides fruits and nuts, but is also a useful shelterbelt, providing overwintering sites for animals and flowers for pollinators in spring.

ORCHARDS

There is so much you can do to improve the health of your fruit trees and the soil beneath them, whether you have a large orchard or just the odd tree. I have a small orchard of 25 trees in the walled garden; it covers an area of 25 × 25m (82 × 82ft) and is a mix of fruit trees (apple, pear, cherry, damson and greengage) grown in rows 5m (16½ft) apart, with the apple trees on M106 rootstocks, so they don't get too large. They were well mulched with compost for the first few years, but it's not been without problems. Trees have suffered bark damage by rabbits and voles, and the two

Plant diversity

'James Grieve' apples suffered from bacterial canker, probably as a consequence of the bark damage, and had to be removed. The groundcover is a mix of grasses, dandelions, buttercups, white dead-nettles, stinging nettles and hogweed, along with some early primroses and other spring flowers that appear before the grass gets underway. We don't cut the grass regularly. We let the trees flower and set fruit before mowing in late spring with a two-wheeled walking tractor which has a sickle-bar mowing head, and we leave the clippings on the surface to rot down. We don't cut under the trees close to the trunks. Depending on the year, we may cut again in late summer and, sometimes, after harvest. This practice helps the soil fungi, giving them plenty of organic matter, and draws in lots of insect pollinators, while the long grasses provide cover for beetles and bark-nibbling voles (of which we have plenty!).

There are other benefits (aside from the saving in labour!) from not cutting the grass too frequently. Every time you cut, you are disturbing beneficial animals and, secondly, long grass makes it more difficult for the spores of apple scab to be blown or splashed from the ground to the newly opened leaves. Interestingly, the tree that suffers most from scab is on a corner adjacent to two paths, and I suspect that the fungal spores get splashed up from the paths. I am more diligent now about picking up the leaves from those paths.

Here are some ideas to improve tree and soil health if you are establishing your own mini orchard:

- Mix up the species and varieties, and maybe include something different, such as a quince or medlar.
- Add more variety with nut trees, such as hazel, and fruiting shrubs like elder.
- If there is space, establish an understory of soft fruit bushes. In one organic orchard I visited, there was a row of cordoned gooseberry bushes between the lines of apples, creating another habitat.

Getting the planting right

- Grow lots of wildflowers, both around and within the orchard. Research shows that up to 18 times more orchard pests are parasitised by wasps where there are flowers, with buttercup and dandelion being particularly useful, along with wild parsnip and carrot.
- Don't mow too often, as long grass gives better natural cover for predators that are attracted by the nectar and pollen of the wildflowers and supports more species that prey on aphids and thrips.
- Growing red clover under fruit trees encourages pollinators and predatory mites.
- Mulch your trees with a fungi-rich woody compost (see 'Bacteria- versus fungi-rich compost', page 28) or willow woodchip. This will boost soil organic matter, increase the populations of beneficial soil microbes, and increase the amount of active soil carbon and nitrogen available to the trees. Mulch the soil under each tree to the drip line (that is, to the edge of the canopy), but don't add too thick a layer or it will kill the groundcover. Also, don't let the mulch touch the trunk. According to Michael Phillips, author of *The Holistic Orchard*, you need about one and a half buckets of mulch per tree.
- Growing a mixed-species hedgerow around an orchard has lots of benefits. In the past, orchards were surrounded by hedgerows, often abundant with fruits such as damsons, which would entice or lure pests away from the main fruit or nut tree crops.
- Plant some rowan trees near your orchard. There is an interesting relationship between the number of rowan berries, apple fruit moths (*Argyresthia conjugella*), whose larvae feed on apple or rowan fruits, and the parasitic wasp (*Microgaster politus*), which is a parasitoid of the larvae. The female moths prefer to lay their eggs on rowan fruits. In a poor

Plant diversity

year, with few rowan berries, the female moths move into orchards and lay their eggs on apples. The larvae damage the apples. Entomologists recommend that rowan trees are planted near orchards so they can provide food for the fruit moth larvae which, in turn, supports more predators. With plenty of predators around, the moths are kept under control and they don't move into the orchards.

REWILDING YOUR GARDEN

I have seen first-hand how the biodiversity of our farm has increased over 16 years through our organic, agroecological management. We've built ponds, planted willow and let our hedgerows grow large and thick. I didn't realise just how important small brambly/scrubby areas were to trees, until I saw how they allow tree seedlings to survive out of reach of browsing animals. Now, we have a new generation of oak trees appearing on the farm.

Rewilding is possible even in the smallest of spaces. It doesn't mean letting nature take over the whole of your garden. If you do, you usually end up with more of the invasive weeds and, while it's fine to have a bramble patch on a farm, you don't want those species taking over a small garden. But you can loosen up and create areas where wildlife can thrive, such as a tiny copse in the corner, an area of long grass, a few weedy patches, some wood piles and small ponds. These will create a mosaic of interconnecting habitats that will attract a wide diversity of species.

The lawn

It's easy to relinquish some control of your lawn. Throw away the weedkiller and fertilisers, and let a few flowers grow. Or simply make a higher cut, so more ground-living animals survive the blades. Even our formal lawns receive no 'treatment', other than an infrequent mow. We collect the clippings as they are full of

Getting the planting right

nutrients and would encourage more dominant grasses (I use the clippings for mulch). Over time, the fertility of the soil has decreased and now favours the less competitive native species. The lawns are now a mix of flowering plants, grasses and mosses that are tolerant of mowing. I have counted 14 different species (other than the grasses and mosses): buttercup, cinquefoil, red and white clover, daisy, dandelion, hawkweed, ground ivy, scarlet pimpernel, plantain, self-heal, sorrel, speedwell and violets. It is fascinating to see how the mix varies from the sunnier side to the shadier east side of the house. Other ideas include creating patterns in the lawn by leaving strips unmown or giving over a strip at the back to create a wildflower meadow, leaving it uncut in summer. You can plug-plant some pretty native meadow flowers to get things going. Cut your meadow once a year, after flowering, to prevent the more aggressive plants from getting established.

Hedges and boundaries

Walk along a hedgerow in late spring and you'll see the open flowers of cow parsley, hogweed and other umbelliferous species covered in predatory wasps and beetles, which then move into the fields to help to control crop pests. The hedgerow provides year-round shelter for natural predators. You can create hedgerows along your boundaries and establish shrubberies elsewhere.

Wildlife shelter

Create shelters and habitats under your trees and shrubs by piling up small logs and covering them with twigs and fallen leaves. These are perfect for ground beetles, spiders and other invertebrates, as well as for snakes, toads and even hedgehogs. You can build bug hotels (see page 77) and wood mould boxes (see page 78). Even piles of old clay pipes can provide an important refuge.

Nettle patch

Nettles (*Urtica dioica*) encourage insects and are the food plant for the caterpillars of the small tortoiseshell, red admiral, peacock

Plant diversity

and comma butterflies, as well as several moths. In fact, around 50 different species of insects are associated with nettles, as the sting protects them from grazing animals. To get the most benefits for wildlife, don't establish your nettle patch in a shady corner as you need a sunny spot to draw in the butterflies. Cut the nettles back in summer to encourage another flush of leaves, after checking first for caterpillars and predators. The nettles can be harvested (wearing gloves) to make nettle tea (see page 147) or to eat, as the young leaves are highly nutritious.

Build a pond

A pond or small wetland area is probably the most beneficial feature you can create for wildlife. It can be as simple as a small container or half a barrel but, if you have the space, having a clay-lined pond with a deep area and a shallow margin is the ideal. It will attract animals such as dragonflies and damselflies, newts, frogs, toads and grass snakes, as well as provide drinking water for birds and mammals.

Use vertical spaces

Make sure walls, fences and arches are covered with plants. A wall or fence covered with a grapevine, honeysuckle or some ivy not only looks wonderful, but can provide nesting sites for many of our garden birds, while flowers and berries are sources of food and attract pollinators. Ivy flowers in autumn provide bees with a last boost before they hibernate for winter.

Plant a tree

There is always a spot for a tree in a garden; you just need to check that it's right for your space. A tree creates another layer in the garden, so it will be colonised by a different group of insects and birds. It will shade the area below, to create more varied microhabitats, and will contribute to the leaf litter in autumn. It will even have a cooling effect on a hot day.

CHAPTER 6

Choosing the right plant

You hear so many garden designers talk about 'planting the right plant in the right place', but it's a piece of common-sense advice that's fundamental to a resilient garden. If a plant is not in the right place, it will be under stress and that means it will be more prone to pests and disease.

Even the experts can get it wrong, though. I have heard many a judge on the Royal Horticulture Society (RHS) Chelsea Flower Show programmes comment on the unsuitability of a plant for a particular position. Of course, they are only temporary show gardens, and the planting density is much higher than you would want in your own garden but, quite rightly, the judges are critical when the designer puts a plant in the wrong place. We all get it wrong, but I think many gardeners are loath to admit their mistake and move a plant once it is in the ground; however, the quicker you move the plant, the more likely you are to be able to give it a new lease of life.

SEED SELECTION

Gardeners have a huge choice of vegetable seeds nowadays, but these have one thing in common – the most valuable attribute seems to be yield. Look at the first line of the description and you will usually see phrases such as 'heavy cropper', 'high yielding' and 'prolific fruits'. Many of these heavy croppers are hybrid crops, known as F1 hybrids – seeds from a cross between two parental plants selected for specific characteristics. I try to avoid these, partly because high yield is not one of my preferred characteristics – I'm more interested in taste and adaptability to my conditions – and because you can't save seed from these plants (the seeds from an F1 hybrid will not breed true; see 'Seed saving – breeding your own resistance', page 110). So, over the years, I try to save as much seed from my own plants as I can, which, year by year, have become more adapted to my conditions.

Commercial seeds

I am particularly concerned about commercial plant-breeding programmes. When breeding primarily for more yield or bigger flowers, the genes that may be involved in natural defence mechanisms may be lost. The result may well be an 'improved' plant that is more productive or has a bigger flower, but it may also be more vulnerable to pests and disease, having lost some of its inbuilt defensive mechanisms, which we know are complex. Much has been written about the loss of scent in some of the improved garden flowers, but the same applies to crops. There are some varieties of cotton that have been bred to be nectar-free, so that they don't attract pests such as moths but, as a result, they don't attract the natural predators either, leaving the plants more vulnerable. Plant breeding tends to look for short-term gains, but with these come unforeseen consequences. Even when the focus of the breeding programme is on a specific plant/pest or disease interaction, selecting for a specific resistant gene can result in the loss of other genes that may be involved with resistance or the attraction of natural predators.

Another problem is that a plant may have resistance against a specific disease, but could easily be susceptible to another disease, so you need to check which ones and make a decision. In the case of blight resistance in potatoes, the pathogen changes regularly, so

Choosing the right plant

within a few years, the disease resistance may fail. In the UK, some of the well-known potato varieties that used to be resistant to blight are now no longer. 'Lady Balfour', long popular with organic growers, is one such example. There are some good examples, though. The Sarvari Trust breeds blight-resistant potatoes under the Sarpo label, and their starting point was naturally disease-resistant tubers that they found growing in different parts of Europe. But don't forget, potatoes are susceptible to a variety of diseases, so consider what is more important on your site – blight, scab or eelworm? Then there are some varieties that have all-round resistance to disease, but do they actually taste good?

Disease resistance and tolerance

Some seed packs describe a plant as disease-tolerant, which basically means that the plants may be able to withstand the pathogen for a while but will succumb if conditions are not ideal. Within a crop, there will be natural variability, with some individuals more able to tolerate the disease than others, or they may have a tolerance of a wider range of diseases than others. A resistant crop has greater tolerance of the disease and is better able to withstand an attack when the environmental conditions are not ideal.

Some crops are naturally less vulnerable to disease or less attractive to pests than others, and they are good choices. For example, there is some evidence that white butterflies prefer to lay their eggs on green-leaved cabbages rather than red, so red cabbages suffer less from damage. The red pigments, which include anthocyanins and carotenoids, make leaves less palatable to caterpillars and even slugs. Similarly, growers find that red-leaved lettuce varieties are less susceptible to other pests, too. It can pay to opt for some of the older, ancestral plant varieties that haven't been 'improved' or even wild species. I get various leaf spots on my chard and spinach, but my sea beet, a wild relative found growing in extreme environments of shingle banks and salt marshes, rarely suffers from disease. Another unusual crop is New Zealand spinach (*Tetragonia tetragonoides*). It looks much like common spinach

Plant diversity

(Spinacia oleracea), but generally suffers from few pests and diseases and has the added benefit of being drought resistant, so it copes well with hot summers during which common spinach tends to bolt. Perennial vegetables, too, seem to suffer from fewer pests and diseases. I rarely spot pests on my skirret (*Sium sisarum*), Good King Henry (*Chenopodium bonus-henricus*) and Chinese artichokes (*Stachys affinis*).

SEED SAVING – BREEDING YOUR OWN RESISTANCE

The natural agriculture approach is to save seed each year (see '3. Continuous cropping', page 97), so that plants become adapted to the local conditions and soil. They also adapt to the lack of inputs, such as fertilisers and pesticides. This is something that I am trying to achieve in my own plot.

F1 seeds are the norm in seed catalogues nowadays. Plant breeders have a number of pure breeding parental strains with desirable characteristics that have been inbred for a number of generations. They take parent plants with desirable characteristics and cross them to produce the F1, the first cross, which inherits characteristics from both parents. F1 seeds are costly to produce, and this is reflected in the fact that they are more expensive than single-species seeds. They are often selected for more vigour and for uniformity, although there may be disease resistance, too. You can't save seed from them, because they do not breed true and the offspring won't necessarily have the same characteristics. They are usually protected by breeder's rights as well.

Open-pollinated seeds

It's very different with open-pollinated seeds. They breed true to type – the offspring will resemble the parent plants and will pass on the characteristics from one generation to the next. They will have a lot of genetic variability, which means that they can adapt to environmental change, and these are plants from which you can

Choosing the right plant

save seed. Open-pollinated seeds are not protected by breeder's rights, nor are they produced from a cross, so the seed is cheaper to buy as well as being available through 'seed swap' clubs.

Sadly, so many of the open-pollinated varieties that were available even 20 years ago are no longer available. It's not because they are no good, but that they are not on the national seed list. This list is maintained by a government agency and if you want to sell a variety of seed, it must be on the list. It's costly to add a new variety to the list, so not surprisingly the focus is on a few varieties that sell well and the F1 hybrids. It's all becoming much more uniform and less varied, and this is designed to ensure that people come back to buy seed each year rather than save their own. This is why it's important to support the smaller companies that save their own seed and often encourage you to save seed yourself. The only point to remember with open-pollinated seeds is that one seed merchant's open-pollinated variety may differ from another, despite having exactly the same name.

How to save seed

Seed saving is really easy and very satisfying. Often, I find that my own saved seed is fresher and of much better quality than that which I can buy commercially, and I find that it produces better-quality seedlings. Getting a good start is important, too, as the seedlings are less likely to succumb to diseases such as damping off.

When saving your own seed, choose your parental plants carefully; you will be growing your next crop from these. Think about the traits that are important to you. It may not be yield or size, but plants that have suffered from less disease, the plants that survived the onslaught of pests and disease, or those that coped with drought or late frost. You don't want to reproduce the less attractive traits, such as bolting, so when saving seed from plants that have a habit of bolting – beetroot, radish and lettuce, for example – don't save seed from those that flower first. Instead, wait and choose parental plants that didn't show any tendency to bolt. Similarly, with leeks and onions, choose the plants that have survived without any rust pustules on their leaves. During the season, it can be helpful to mark the plants from which you wish to save seed, either with a tag or a cane.

If you are intending to save seed, always grow more plants than you need, so you can save seed from about 10 per cent of your crop. Annuals are easier, as you save seed at the end of the growing season. Biennials, such as beetroots, carrot, celeriac and chard, are a bit trickier, as they flower in their second year. You can either leave the plants to overwinter where they are or, better still, lift and store them over winter. Lifting means that you can check the roots. Simply lever them out of the ground, taking care not to damage the roots (especially those of parsnips), twist off the leaves and store them, carefully labelled, in a cold room over winter. They can be replanted in early spring, to resume their growth, flower and set seed. Onions can also be lifted, stored and replanted in early spring. Another advantage with lifting is that you can move them to a more convenient bed well away from other crops of the same type or you can replant them in a polytunnel. Some crops, such as brassicas, will be harvested over winter and early spring, so once you have finished taking a crop, let them flower.

MY OWN BREEDING PROGRAMME

I am trying out a bit of plant breeding myself. I have saved parsnip seed for the last 10 years, with the open-pollinated 'Tender and True' as the parent generation, a heritage variety chosen for its resistance to canker. Like those practising natural agriculture, my aim is to breed a parsnip that is adapted to the conditions under which I grow it, as I harvest seed from healthy plants that grow well on my clay soil.

I've been researching and writing about climate-change gardening for a while, so I always think about the conditions that I might be facing in coming years. One question that I ask myself is whether my saved seed will be right for my future garden. 'Tender and True' dates back to 1897, when it was developed by Victorian gardeners under very different climatic conditions. These were times when the average temperatures were at least 1°C (34°F) lower and when cold winters and extended periods of below-freezing temperatures were commonplace. With the pattern of today's rainfall predicted to change from an all-year-round distribution to one in which most of the rain falls in winter, together with less frequent but intense periods of severe rainfall and drought, should I be looking for a parsnip that can cope with low summer rainfall and potentially higher temperatures and even drought?

Population wheat

I have been following an exciting wheat-breeding programme called 'ORC Wakelyns Population'. It's not a single-variety wheat, but the result of the careful crossbreeding of several varieties, which were then allowed to cross again and again. A field of population wheat looks very different to one of a modern variety. It's not uniform, but varied; some stalks are tall, some short, some have heavy ears, others only small ones; and the plants will vary in disease resistance. The theory behind population wheat is that some of the plants will thrive, while others won't, which means that yields won't be the highest. However, when disease turns up, while some of the plants may succumb, the crop won't fail completely. It's a great idea that's gaining traction because it is helping farmers to adapt more quickly to climate change.

The key to its success is repeated seed saving. You don't want to keep breeding the same plants as conditions change. If a new strain of yellow rust emerges, you want to select seeds from plants with more resistance. But, in a few years, you may be selecting for a different strain or even a different disease. You really want your wheat population to be dynamic and adaptable. And the idea of populations is spreading to population rye and others.

So, now, I'm trying it with my parsnips and have just finished year two. The first year, I grew two varieties, my own and 'Parsnip Guernsey'. I let the best plants overwinter, flower and cross-pollinate the following year (parsnips are biennials) and I saved

Plant diversity

seed from these plants. Now I have plenty of saved seed and, each year going forwards, I'll choose the best plants to be parents. If I have a few bad years, I may introduce another variety.

I'm introducing more diversity into other crops, too, by sowing a mix of varieties rather than just one, in order to hedge my bets. If disease or conditions result in the failure of one variety, the other varieties may thrive. I may lose some but, hopefully, never the whole crop. When it comes to seed saving, it's going to be a mixture, just like population wheat.

GRAFTING

Grafting is a traditional method used to help control soilborne disease, improve stress tolerance and yield. It's a process that involves attaching a cutting (called a scion) from a donor tree to a rootstock, so that the two parts join together to form one plant. The rootstock determines the vigour and size of the tree and, to some extent, the disease resistance, whereas the cutting defines the variety of the fruit.

Fruit trees have traditionally been grafted. When you buy a fruit tree, you choose a variety and then a rootstock, each according to your needs. Both rootstock and variety can differ in their resistance to disease, with some apple varieties being less susceptible to apple scab or bacterial canker. For apple trees, the rootstocks range from M25, which is very vigorous and produces a standard tree that grows to 4m (13ft) and more, to the popular MM106 which is a half-standard and is suitable for espaliers and cordons, to the M26 dwarfing rootstock which is suitable for growing in containers. There are also rootstocks for plums, pears, gages and cherries. Careful choice of variety and rootstock can help avoid some of the more common fruit tree diseases.

Grafted vegetable plants were once only available to commercial growers, but now gardeners can buy grafted tomatoes, cucumbers, aubergines and more besides. They are more costly because of the labour involved in carrying out the grafting, but

Choosing the right plant

there are definite benefits in terms of yield and disease resistance. The grafted plant consists of a scion with desirable fruit characteristics and a rootstock that may give disease resistance, vigour, yield and drought tolerance. It is important to look at the characteristics of the grafted plant carefully and choose characteristics that suit your particular growing conditions.

Disease-resistant varieties and rootstocks are a good step forward but, for the long term, I think you need to consider the whole of the plot and the overall level of disease – taking a more holistic approach, focusing on healthy soils and saving your own seed.

PART 4 Boosting defences

CHAPTER 7

Biocontrol

There are times when, despite all your best efforts, you get a pest problem that you can't control through natural management. Walking around my polytunnel in summer, my heart sinks when I see my cucumbers suffering from mottled leaves. It could be a nutrient deficiency, so I'll give them a dose of comfrey tea or seaweed feed, just in case, but it's more likely to be a red spider mite infestation. The polytunnel is one of two areas where I'm least successful at controlling pests and disease, the other being a narrow, sunny Victorian conservatory along the front of the house which is home to an ancient vine. Both have incredibly high levels of heat and humidity in the spring and summer months, and this creates perfect conditions for pests to thrive. Despite my efforts, red spider mite and other pests recur almost every year and, to tackle this problem, I resort to biocontrol.

HOW DOES IT WORK?

Biological control, or biocontrol, is the use of one organism to control a pest or disease-causing species. Ladybirds, lacewings, beetles and parasitic wasps are just a few of the many natural predators that provide our growing spaces with natural pest control. Biocontrol is the extension of the idea of the natural predator. However, rather than waiting for natural predators to arrive and provide control naturally, predators are introduced into a system artificially, so that they can act more quickly. This may seem like a relatively new idea, but it has been around for about 150 years. The first example of its kind dates back to 1868, when the newly established citrus groves in California suffered infestations of the cottony cushion scale insect (*Icerya purchasi*), a leaf-sucking insect that caused defoliation and dieback of the branches. It had a devastating impact on the young citrus groves and there was no known treatment until a natural predator, the vedalia ladybird beetle (*Rodolia cardinalis*), was discovered in Australia and introduced to the citrus groves.

The growers were lucky that the biocontrol worked, as there have been plenty of failed examples in which the predator has become a massive problem, such as the release of cane toads (*Rhinella marina*, formerly *Bufo marinus*) in 1935 to control the greyback cane beetle (*Dermolepida albohirtum*), a pest of sugar cane in Australia. Once introduced, the cane toads soon moved off to pastures new, eating almost anything they could find on the ground, including lizards, snakes, frogs and other small animals, and outcompeting native species. Despite campaigns to eliminate the cane toad, it now has a widespread distribution and continues to threaten the survival of many native species. Fortunately, this lesson seems to have been learned. Nowadays detailed risk assessments are carried out before predators are introduced and most biocontrols tend to be based on a single crop and pest.

The range of species available to purchase for biocontrol is increasing each year. Some are soil-based, while others are for use in a controlled environment, such as a glasshouse or polytunnel. In this chapter, I'll cover the most common beneficial species that you can buy for biocontrol.

WHY USE BIOCONTROL?

There are benefits to using biocontrol rather than other methods, even relatively benign ones. For example, many gardeners consider it safe to spray soapy water over aphids, but even this method

Biocontrol

can have consequences for non-target species, such as ladybird and lacewing larvae.

Pests can soon develop resistance to pesticides, so biocontrol is often the preferred means of control and, once established, the control can last for much of the growing season. As you will see, the beneficial organisms used in biocontrol have specific requirements, especially with regard to temperature. So, it is important to make sure the environment is right before you introduce them. Also, beneficial organisms can take a while to get established, so they don't offer instant control, but, being specific to a particular pest, they won't harm non-target organisms. However, you need to be observant. If you see a problem developing, order the biocontrol rather than wait. But, equally, don't order and apply it too soon because the beneficial organism needs food! If there are too few pests around, they will not thrive and reproduce, and you will end up ordering again later in the season. Timing is everything.

As I mentioned earlier, my conservatory and polytunnel are the two places that I find it hardest to control pests. In addition to the hot, humid environment that encourages pests and disease, the fact that you are always harvesting crops and then replanting means there is little opportunity for a balanced ecosystem to build up. It's a very stressful environment for plants, given the temperature, irrigation and the density of the plants, so not surprisingly, some pests multiply quickly under these conditions. Many of the biocontrol treatments available on the market have been developed for these controlled environments. They tend to require warm conditions, usually a daytime temperature of 17–20°C (63–68°F) in order to grow and reproduce quickly.

NEMATODES

Among the many different types of nematode are the predatory ones that feed on specific species, even other nematodes. Each has its own mutualistic bacteria living in its gut. Those chosen for biocontrol enter the pest's body through an opening and then

Boosting defences

release their bacteria. It is the bacteria that digest the pest and cause it to die. Once the pest is dead, the nematodes then feed on the remains. Interestingly, an expert can tell what species of nematode has been active from the colour of the remains, which range from bright yellow to red.

A range of different nematode treatments has been developed, each using a specific species of nematode to control a particular pest. For example, nematodes can be used to control slugs, vine weevils, carrot root fly, cabbage root fly, cutworms, onion fly, leatherjackets, sciarid fly, box tree caterpillars, gooseberry sawfly, thrips and codling moth.

Most nematodes are supplied as a powder-like grow media in trays with a micro-perforated film lid to allow air in and keep them alive. A living product, they must be used immediately, though they can be stored in a fridge for up to 14 days. When you are ready to use them, the contents of the tray are mixed with water and applied as a drench using a watering can with a coarse rose attachment or a hose end feeder. For the treatment to be successful, the ground temperature must be above 5°C (41°F), even at night, and the soil must be moist. You may have to water the soil to keep it moist and allow the nematodes to thrive. It takes several days for the nematodes to become active and several weeks for their effects to be detected. Very often, you won't see any obvious results as there are no dead bodies lying on the soil surface!

There is a new nematode treatment on the market, developed in North America, which avoids the need to make up a drench. Instead, the nematodes are provided in a granular form and then sprinkled around plants that need protecting. They become active when they are sprayed with water. The granules protect the nematodes in the soil for the next 5 to 10 days, preventing them from dehydrating before they make contact with the pest. This treatment can be used for slugs, fungus gnats, thrips, vine weevils and grubs, so it makes a useful alternative to the nematode drench.

One of the most widely used nematode treatments is that for slugs. There are many ways to control slugs, but in years where damp conditions favour their growth and reproduction, they can be almost impossible to control. One particular nematode, *Phasmarhabditis hermaphrodita*, is a specialist slug hunter. Juvenile nematodes actively seek out slugs and, when they find one, enter its body through the breathing pore. Once inside, they release their payload of bacteria which attack the slug. The slug stops feeding and retreats underground to die. Nematode treatments work for up to six weeks, so if you have a particularly slug-ridden area, you may want to apply the treatment at six-weekly intervals from early spring to early autumn, when temperatures are above 5°C (41°F) and slugs are active.

USING PARASITIC WASPS

Polytunnels and greenhouses, with their higher temperatures and humid air, are very likely to suffer from attacks of whitefly (*Trialeurodes vaporariorum*) and aphids. You know you have whitefly when you shake a plant and a mass of small white specks, the flies, rise up from the leaves. Whitefly become active as soon as temperatures start to warm up and just when you are thinking of planting out your early season crops! I usually get them on my early courgettes. I have controlled whitefly by washing them off with a hose and removing badly infested leaves. But, if all else fails, I have resorted to buying a vial of small parasitic wasps called *Encarsia formosa*. The wasps are supplied on white cards which are hung on the lower stems and branches of plants. They have to be placed near to the whitefly, as the tiny wasp is a weak flier. On finding a whitefly, the female wasp inserts its ovipositor into its body and lays an egg. The resulting larva eats the whitefly from the inside. You know the treatment has worked when the whitefly turn black. A single female wasp can lay as many as 300 eggs, which means you can assemble a formidable army of wasps in just a few weeks.

There are many different species of parasitic wasp. If the problem is aphids, choose either *Aphidius colemani* or *Aphidius ervi*. Unlike *Encarsia* wasps, these are supplied at the mummified

Boosting defences

Table 4. Beneficial nematodes

PEST	NEMATODE
Ants	*Steinernema feltiae*
Caterpillars (including gooseberry sawfly, box tree moth, cabbage white)	Generic mix
Chafer grubs	*Heterorhabditis bacteriophora*
Codling moth caterpillar	Generic mix
Leatherjackets	*Steinernema feltiae*

Biocontrol

TREATMENT

Nematodes do not kill ants; instead, the parasitic nematodes drive the ants away from the treated area. Apply the nematode drench around the roots of the plant affected by ants at any time of year.

When treating caterpillars, the nematodes need to be sprayed directly onto the pest. This generic mix of nematodes is applied when temperatures during the day are at least 12°C (54°F). To treat gooseberry sawfly, apply the drench to the leaves as soon as the first caterpillars are spotted. You may need to repeat at weekly intervals. Similarly, with box tree moth and cabbage white caterpillars, use when you see the caterpillars between early spring and mid-autumn.

Chafer grubs live in the soil under grass, feeding on grass roots and creating yellow patches. The nematode drench is applied in either autumn or late spring when the grubs are near the surface and the soil temperature is about 12°C (54°F). The rest of the year, they are deep in the soil or have pupated. The nematodes enter the body of the grub via body openings and release bacteria that kill the grub. The nematodes reproduce inside the dead grub and release more infective nematodes. If there are plenty of grubs, the nematodes will increase in number and can exert control for several years.

Codling moth caterpillars damage apples and pears by burrowing into the flesh. Once they have fed, they drop onto the bark of the tree or the soil to overwinter. To treat, apply the nematode drench to the tree trunk and underlying soil in early and mid-autumn. Give three treatments at two-weekly intervals.

Leatherjackets are the larvae of the European crane fly (*Tipula paludosa*). The adults appear in late summer and lay eggs. The young larvae are most sensitive to the nematodes, so apply the drench two weeks after you see the adults. The soil temperature needs to be 10–30°C (50–86°F) during treatment and for the next two weeks. Cut the grass short first, apply the drench and water to

Table 4. Beneficial nematodes (*continued*)

PEST	NEMATODE
Onion fly	Generic mix
Scale insects	Generic mix
Sciarid fly (fungus gnats)	Generic mix
Slugs	*Phasmarhabditis hermaphrodita*
Vine weevil	*Steinernema kraussei*

stage – that is, a dead aphid that has been 'mummified' by a parasitic wasp larva, which is inside and ready to metamorphose into an adult wasp. Unlike *Encarsia* wasps, these wasps will fly away, so you need to cover doors and other openings with screens. Another option is a predatory gall-midge, *Aphidoletes aphidomyza*, the

Biocontrol

> **TREATMENT**
>
> wash the nematodes into the soil where the larvae are feeding. The soil must not be allowed to dry out for two weeks. You can re-treat in spring, but the mature leatherjackets are more difficult to kill this way.
>
> Onion flies are active all summer, laying their eggs near the base of plants. The larvae are found in the stem and leaves. There are several generations, with the pupae of the last generation overwintering in the soil. The drench is applied to the soil around the plants and repeated at 14-day intervals.
>
> Once scale insects are spotted, spray the leaves. The temperature needs to be a minimum of 14°C (57°F). Apply at three-weekly intervals. The scales do not drop, so they need to be rubbed off.
>
> Once the flies are spotted, drench the soil around the plants, including potted plants, between mid-spring and mid-summer.
>
> Slugs have a peak egg-laying period of early to mid-spring and early to mid-autumn. Apply as a drench when the soil temperature reaches 5°C (41°F), although they can survive the occasional frost. Apply every six weeks. Potatoes should be drenched six weeks before harvest to reduce damage to their tubers.
>
> Vine weevil larvae feed on the roots and stem bases of plants from mid-summer to winter, weakening and often killing the plant. The problem tends to be most severe in container-grown plants. The nematode treatment is most effective in late summer to autumn with a follow-up application in spring.

larvae of which hunt and kill aphids. And finally, you could buy larvae of the lacewing (*Chrysopa* spp.) and two-spot ladybird (*Adalia bipunctata*).

Parasitic wasps can also be used against scale insects. These pests are commonly found on citrus fruits, camellias and conservatory plants. They hardly move and attach themselves firmly to the stem and underside of leaves where they can suck sap. This results in yellow leaves and, sometimes, defoliation. They also produce honeydew which drops onto the leaves and is colonised by a sooty mould, turning everything black. The female scale insect lays hundreds of eggs which remain protected under her scale until they hatch. The small, crawling larvae then emerge and move to another part of the plant to continue their life cycle.

The tiny yellow, parasitic wasp *Metaphycus helvolus* parasitises the young stages of the scale insect when they have just settled on the plant. The wasp lays its eggs in the soft scale and the larvae develop inside and kill it. Parasitised scales look darker and flatter than unaffected ones. Adult wasps are long-lived, surviving for two months if they have a good supply of food. However, like most parasitic wasp treatments, they must be used in a contained space and require a minimum temperature of 22°C (72°F) for a few hours a day and good light levels. They can be used from late spring through the summer. If you have scale insects, another treatment option is nematodes (see 'Table 4. Beneficial nematodes', page 122).

PREDATORY MITES

The single biggest problem in my polytunnel is the red or two-spotted spider mite (*Tetranychus urticae*). I first knew I had problems when I noticed that some of my overwintering spinach plants were covered in a fine web. On looking closely, I could see that the plants were alive with small, moving dots – red spider mites. I removed all the infested plants and leaf debris and hoped I had controlled the problem. But, within a week or so, I noticed that my cucumbers were suffering from bleached leaves and dying back. This was also classic mite damage. Typically, red spider mites are found on the underside of leaves where they suck sap, leaving distinctive white speckles. Eventually, the leaf becomes bleached and dies.

Red spider mites overwinter in crevices and plant debris, emerging in spring to feed on plants. They reproduce at an alarming rate, so

Biocontrol

a few can become thousands in just a few days. In really bad infestations, you see webs around the growing tips of plants, just as I did. This pest prefers warm, dry conditions, so increasing the humidity can help to control the number. But this rarely gives complete control, so I opted for a predatory mite called *Phytoseiulus persimilis*. These active, fast-moving mites have a voracious appetite, eating up to five adults or 20 eggs a day, but this species can only be introduced if you already have plenty of spider mites as they need lots of food! *Phytoseiulus* mites are sent in a vial and you simply tip them out onto your plants, where they get to work. It takes several weeks for the mites to gain control of the problem. However, this species needs warm temperatures, so if the temperature is likely to fall below 10°C (50°F), you will need to use another species. *Amblyseius* can also control thrips. It is a better choice for low-level infestations and the cooler conditions that exist earlier in the growing season.

It can often be difficult to know whether biocontrol has worked. For me, within three weeks it was obvious that the cucumbers had overcome their crisis, as they were putting on new, healthy growth. However, I now assume that there will always be some overwintering red mites, since biocontrol doesn't wipe the pests out completely. I use a lot of mulch so, at the end of the year, there is a lot of organic debris on the soil which provides perfect hiding places for red spider mites. Also, I don't water in winter unless I am growing some crops and, annoyingly, the dry conditions favour the red spider mites, too. *Phytoseiulus* won't survive the winter, so when the temperature warms up, the red spider mites will not have any predators and their numbers can increase quickly, so I now apply *Amblyseius* early in the season to make sure that the spider mites do not have a chance to thrive.

There are a couple of other options that are particularly useful early and late in the season. A predatory gall-midge called *Feltiella acarisuga*, which is active in spring and autumn, can be released to prey on small populations of red spider mite. The females hunt out clusters of red spider mites and lay their eggs nearby. The yellow larvae hatch and eat the mites. Another option is the rove beetle. I've mentioned rove beetles before, but there is a species that you can introduce specifically, the greenhouse rove beetle (*Atheta coriaria*), which hunts thrips, fungus gnats and red spider mites.

Predatory mites can also be used against other pests. Tiny flies called sciarid flies or fungus gnats are common in greenhouses and polytunnels. The adult flies are found on the surface of the soil and around leaves while their white larvae live in the soil where they damage seedlings and cuttings. Long-term control can be achieved through the use of tiny, soil-living mites called *Hypoaspis* which are supplied in vials and can be sprinkled onto plants. These mites can survive for several weeks without food and can be released as either a preventative or a control treatment. They are active at temperatures of 16–39°C (61–102°F), and require a minimum temperature of 10°C (50°F). The alternative to mites is nematodes (see 'Table 4. Beneficial nematodes', page 122). You can also buy the predatory mite *Amblyseius andersoni* to combat fuchsia gall mites.

PREDATORY LADYBIRDS

My ancient grapevine inside a Victorian conservatory has had a mealybug infestation for quite a while. Mealybugs are tricky pests to control, as they creep into inaccessible places and cover themselves with a waxy secretion. They feed on sap and excrete a honeydew all over the leaves, branches and fruits which, in turn, creates conditions perfect for sooty moulds. Biocontrol is the answer. The predator for mealybugs is *Cryptolaemus montrouzieri*, a predatory ladybird from Australia. These ladybirds need high temperatures to be active, in the mid- to upper 20°Cs (70°Fs), so control can be achieved only during the summer months. You buy the ladybirds as larvae, which start eating the mealybugs immediately. Then they pupate and become adults. The adults lay their eggs in the egg masses of mealybugs and, as they can each lay up to 500 eggs, they need a good supply of mealybug eggs! The downside of these predators is the fact that they need to be contained or they will fly away, so doors and vents need to be screened to prevent their escape. An option for the citrus mealybug is *Leptomastix dactylopii*, a small parasitic wasp that also needs high temperatures.

TRICHOGRAMMA AND OTHER WASPS

Parasitic *Trichogramma* wasps are the most widely released biocontrols. The tiny female wasp lays her eggs inside a recently laid egg of the host. The egg gets blacker as the larva develops. These wasps parasitise a wide range of insects, including moths, butterflies, beetles, flies, wasps and bugs. *Trichogramma* wasps are widely used for biocontrol in North America, where cards loaded with the parasitised eggs of a non-host species are placed in fields to control pests such as the cotton bollworm (*Helicoverpa armigera*) and European corn borer (*Ostrinia nubilalis*). In the UK, these wasps are used against the clothes moth (*Tineola bisselliella*).

MINUTE PIRATE BUGS

Minute pirate bugs (*Orius* spp.) are tiny, fast-moving predators with a voracious appetite for thrips. They are native to North America, where they are used as a biocontrol against thrips in a wide range of crops, including aubergines and sweet peppers, and also against the red or two-spotted spider mite (*Tetranychus urticae*) on beans.

FUTURE DEVELOPMENTS

As biologists learn more about the phyllosphere (see page 131), they are looking for bacteria and fungi from this microhabitat that they can use against plant diseases and pests. A bacterium called *Pseudomonas syringae* is commonly found in the phyllosphere, where it produces a variety of bioactive compounds that act against plant pathogens. There is a new commercial product called

Boosting defences

Biosave which contains a strain of this bacterium. It's used against a number of fungi that damage crops post harvest.

While I am a fan of biocontrols in certain circumstances, I don't recommend using them all the time. Try to restrict their use to when you have a problem and make sure you have identified the pest correctly. Also, check the requirements of the beneficial species, because there is no point ordering them if the temperature is too cold, for example. Ideally, you should be working towards a system in which the natural predators are present in sufficient numbers to provide effective control. It is only when natural predators can't keep a lid on the pests or when extreme weather events, such as a heatwave, have put the natural ecosystem out of balance, that biocontrol systems come into their own. Once you have got the pest under control, take a look at your management and see how you could make changes to favour natural predators and put the pests at a disadvantage.

CHAPTER 8

Plant defences

A healthy plant is a resilient plant, as it is best able to withstand the onslaught of a pest or disease. This is achieved by having a good soil and the right conditions for the plant – so that it is not stressed. Another way to help your plants combat pests and disease and even cope with extreme conditions is by boosting their own protective systems.

THE PHYLLOSPHERE

One of the most important interfaces for a plant is the phyllosphere – all the plant's aerial surfaces where they come into contact with microorganisms and the atmosphere. But despite its importance, this microhabitat has hardly been studied. It plays a key role in the health of plants and how they withstand the challenges of pests and pathogens. The interactions that take place between the plant and the microbes in the phyllosphere and between the phyllosphere and animals are incredibly complex.

The microbial inhabitants of the phyllosphere are mostly bacteria, but there are algae, fungi and yeasts, too. Research shows that there can be as many as 10 million bacteria living on the surface of a leaf and they are different from those living in the soil. This microcosm of life is truly ephemeral, because it varies during the day as the microclimate is altered by changes in temperature and light levels, wind speed and humidity. When it rains, millions of bacteria can be washed off the leaves onto the ground.

The composition of the phyllosphere differs between plant species. Different species are host to different types and numbers of bacteria, with more found on broad-leaved plants compared with narrow-leaved grasses and cereals. The composition varies through the growing season, too, with young leaves having mostly bacteria. As the leaves mature, they are colonised by more yeasts and, as they approach senescence, filamentous fungi take over, so it's a kind of 'mini-succession'.

Even more intriguing is the way the community of microbes in the phyllosphere can affect a plant's health and development by supplying biofertilisers, biostimulants and biopesticides. Many of the microbes protect the plant directly. For example, some caterpillars show reduced growth when they eat a leaf infected with beneficial bacteria, while other bacteria can suppress disease-causing organisms. No wonder fungicides and pesticides have a long-term harmful effect; not only do they kill the target organism, but they probably wipe out much of this valuable community, too. It's a truly fascinating world that we are only beginning to discover and we can use this knowledge to boost plant health.

SELF-DEFENCE

Plants can't escape from a pest, so they have evolved sophisticated surveillance systems and defence mechanisms to help protect themselves and even to warn their neighbours and attract help from natural predators.

Plants have an arsenal of defences. There are structural defences, of course, such as thorns and hairs. For example, hairs found on the surface of bean leaves impale aphids and leafhoppers. Of more significance, however, are a plant's chemical defences. When wounded, pine trees produce resin to trap the pest; oak leaves are filled with unpalatable tannins, while pelargoniums release a toxin that paralyses insects for many hours, leaving them defenceless. Some plants stockpile these toxins so that they can make an immediate response to insects, while others manufacture them only when they are attacked by a pest. On detecting an attack, a plant's first response is to strengthen the cell walls in the area under attack, often by laying down more lignin so the tissues are less palatable to pests. Pathogens are prevented from spreading within the plant by deposits of callose. These responses are coordinated by a sophisticated network of pathways around the plant under the control of hormones.

Chemical warfare

Plants release a wide range of chemicals as scents which diffuse through the air. These are referred to as volatiles because they evaporate easily when exposed to air – the scent of a flower, the smell of crushed mint or cut grass are all volatiles. They are produced by all parts of the plant and, while some are used to attract pollinators, others play a role in defence. They are used by plants to communicate, both internally and with other plants.

Plants have a wide range of volatiles at their disposal. Some volatiles repel pests, while others act as an alarm call to bring natural predators to their aid and eat the pest. They can be complex mixes of compounds which differ according to whether the plant has been mechanically wounded, say by a gardener, or if it is under attack from a pest. For example, when a plant is wounded by a leaf-eating caterpillar, it can detect chemicals that are present in the saliva of the chewing insect and release volatiles to repel the insect and attract predators such as parasitic wasps that recognise these signals and come to the plant's aid by laying their eggs in the caterpillar. Messages are sent to the rest of the plant to beef up the production of protective chemicals in the undamaged parts of the plant. The plant may even induce a defensive response in neighbouring plants. It's been found that microbes, both in the phyllosphere and rhizosphere, help in these defences. For example, when maize roots are under attack from a larval pest, they release volatiles to attract a nematode called *Heterorhabditis megidis* which predates the larvae, while apple trees produce volatiles to attract predatory mites to attack spider mites. Mycorrhizal fungi have a role to play, too. They have been found to make resources available to crop plants that can be used to manufacture defensive materials and reduce the level of pest activity as well as to help the plant to withstand pest attacks.

It can be true 'chemical warfare' when the plant produces one compound and the pest then counters it with another. A plant may release natural insecticides, but the attacking insect is equally resourceful and produces its own chemicals. Beetles are expert at this. When they detect the plant's insecticides, they release chemicals that are resistant to these defences. Sometimes, releasing volatiles works against the plant, as it can be a signal to other beetles, telling them that the plant is under attack – this has been found to be the case for the Colorado beetle (*Leptinotarsa decemlineata*) that attacks potatoes. Some beetles can even evade plant defences by working with bacteria in such a way that the plant recognises them as bacteria rather than beetles and doesn't react – a form of 'chemical camouflage'.

Supporting plant defences

How can a gardener help plants that are under attack? The key is having a healthy population of natural predators that can come to the plant's defence, so make sure you have the right flowers and overwintering spots. This will encourage the predators to stay in your garden. It's also useful to grow plants that are rich in essential oils. Not only do they have good defences themselves, but we can harvest plants such as basil, mint, oregano, rosemary and thyme to make insect-repelling sprays (see 'Essential oils', page 148).

Should we be trying to trigger plant defences ourselves? My first instinct when I see a caterpillar is to remove it. Should I remove all the caterpillars? Perhaps it's better to leave one or two, so that they trigger the plant's defences and alert other plants in the area that caterpillars are around? Maybe we need to tolerate a few pests in order to ensure a protective response by our plants, rather than spraying or picking them all off and leaving the plants in an 'unwarned' state.

COMPOST TEAS AND FOLIAR SPRAYS

I know many gardeners pop comfrey or nettle leaves in a bucket of water to rot down and make an effective liquid fertiliser. I do this myself, and I have buckets of comfrey leaves in preparation all summer, taking advantage of the fact that comfrey is a deep-rooting plant and that its leaves are rich in nutrients and minerals. A compost tea, however, is different. It's an aerated mix of good-quality compost and rainwater, and the product is a liquid that is rich in nutrients and contains a diverse mix of microbes. There is mounting evidence that compost teas can improve the health of soils and plants, and indeed farmers who have used large-scale compost tea brewers to spray their arable fields report that they have seen an increase in soil fungi.

Spraying compost tea over soil boosts the number of beneficial fungi and bacteria and also reduces the risk of soilborne pathogens. Growers find it more effective to drench the soil after sowing a crop, and then repeat the treatment while the crop is still at seedling stage. Spraying compost tea over the leaves as a foliar spray helps to reduce plant stress and also acts like a biofungicide. It reduces the incidence of disease by competing with the pathogens on the leaf. Growers have reported that foliar spray of compost tea has been found to be effective against potato and tomato blight, powdery mildews and grey mould (*Botrytis*) on strawberries.

Making your own compost tea

This is relatively easy to make, and you can even buy compost tea kits online. I have a large white bucket and a small aquarium aerator brick. I add a handful of sieved compost and a couple of tablespoons of molasses. Then I fill the bucket three-quarters full with rainwater (no chlorine) and leave to aerate for 24–48 hours.

Boosting defences

Molasses (choose the less refined blackstrap molasses sold in agricultural merchants) are nutrient-rich, encourage the growth of microorganisms and help the compost tea adhere to leaves. Then I strain the tea through a muslin-lined sieve and use a watering can in order to deliver a gentle spray over my crops, particularly seedlings. You don't want to apply it with a high-pressure spray, as this will kill the fungi in the mix. The residue can go back in the compost heap or on the flower beds.

The final tea will be influenced by the ingredients that were used to make the compost. If there was woodchip and lots of brown material, the compost tea will be rich in fungi and ideal for use around perennial plants, while a tea that had more in the way of green materials and farmyard manure will be bacterially rich and great for crops.

If you have a healthy soil, rich in microbes already, I don't think there is much to be gained from using compost teas. The time to use them is when you are working to improve your soils and build up soil life.

When to use as a foliar spray

If you have an outbreak of fungal diseases, such as mildew, on the leaves of a plant or you have problems on vines or fruit trees, then you may see some benefit from a foliar spray.

Some growers report that spraying their crops with seaweed extract is effective. Seaweed extract is rich in a plant hormone called cytokinin. Cytokinin boosts plant growth and seed and fruit production, and it can also delay senescence. In addition, high levels of cytokinin have been found to increase a plant's defences. When a plant is stressed, the cytokinin levels fall and this reduces its defences against pests, so giving a foliar spray of seaweed can help the plant by boosting its resistance. Cytokinin has been found to be produced by bacteria, fungi and algae and can induce plant resistance against pathogens, so that's another good reason to use foliar sprays.

Plant defences

BIOSTIMULANTS

There is an emerging field of plant health products called biostimulants that don't affect the pests and pathogens, but boost the plants own defences (and there are EU and UK regulations to ensure they do as they claim). Some are simply a mix of nutrients, while others are a mix of microorganisms, rather like compost teas. Research shows they work best for a plant that has been stressed. We still don't know enough about the way these products interact with plants, but applications of the right biostimulant can boost shoot and root growth.

Timing is critical, as the application of a biostimulant can have different effects depending on the growth phase of the plant. Take potatoes, for example, where there is a biostimulant that is a polypeptide (a short chain of amino acids). If applied early in the season, the biostimulant increases the number of tubers because it reduces apical dominance and allows more eyes to develop on the seed tuber. When applied mid-season, it helps the crop cope with water stress, and when applied later in the season, the tubers develop a thicker skin which is more resistant to blight fungi and they store better.

There are lots of different biostimulants on the market, including combinations of beneficial bacteria and fungi that mimic the action of soil life in protecting plant roots against chewing insect pests. In particular, there is a bacterium that targets these insects by damaging the chitin which makes up their exoskeleton; there is a seaweed extract rich in macro- and micro-nutrients that boosts rooting and general growth (cytokinins); there are mixes of humic and fulvic acids that bind with nutrients in the soil to make them more available to plant roots; and, finally, there is a powder made from ground volcanic rock which is rich in minerals that boost soil life. Do you need to use them? If you have a healthy soil, then I don't think so, but, as for compost teas, if your soil is damaged or in need of improvement, then these could help.

BIOFUNGICIDES

Similar to biostimulants, biofungicides are formed from beneficial fungi and bacteria found naturally in soil. They work directly against soilborne pathogens such as *Microdochium* (formerly known as *Fusarium*), *Phytophthora* and *Rhizoctonia*. They can produce antimicrobials and even physically shield the plant from the pathogen by forming a protective sheath around the roots. They work indirectly by kickstarting the plant's own defences, and they have been found to boost plant growth by working with mycorrhizal fungi in the soil. However, biofungicides are not treatments but preventatives, so they have to be applied before the disease has struck and must be repeated through the season.

There are numerous biofungicide formulations, each intended to be used on specific crops against particular diseases. One such product is Cerevisane, a purified extract of yeast. It's being used in glasshouses to control moulds on aubergines, tomatoes and strawberries, mildew on cucumbers, and downy mildew on lettuce, while outdoors it can be sprayed on grapevines to control fungal diseases, too. It works by mimicking a fungal attack, activating signals that trigger defences. This results in the production of antimicrobial products, as well as the strengthening of cell walls and the thickening of leaf cuticles. Products containing Cerevisane are available to commercial growers and I am sure it's only a matter of time before they are adapted for gardeners to use.

VACCINATING TREES

I have known for many years that you can trigger plant responses by using salicylic acid (aspirin), and I like to treat my tomatoes with a dose. I dissolve half an aspirin in a few litres of water and feed it to my tomato plants towards the end of the fruiting period, tricking them into thinking they are under attack from pests or disease. They respond by pumping sugars into their fruits to make them sweeter and more attractive to the animals that will disperse

Plant defences

the seeds and so ensure their survival. Vaccination works in a similar way, triggering plant defences and boosting immunity levels.

With climate change and the problems with global plant security, arboriculturalists are having to cope with ever more pests and diseases, and the situation is only going to get worse. In fact, Dr Glynn Percival of the University of Reading is quite downbeat about the future of oak, ash and horse chestnut trees in the UK because of the threat of disease.

Currently, we treat tree disease by either spraying a fungicide, which is not easy given the size of trees and the cost of the treatment, or by felling the affected tree and burning the wood. This latter approach doesn't give the species any time to respond to the disease and build up its own defences to fight off the pest. Remember that trees react very slowly, so a response could take many years.

So, why not take a vaccination approach? Interestingly, it's not a new idea. More than 100 hundred years ago, scientists inoculated single leaves of tobacco plants with tobacco mosaic virus to reduce the levels of subsequent infections with mixed success. More recently, researchers in the Netherlands have vaccinated more than 500,000 Dutch elms with *Verticillium albo-atrum*, a fungus that has been found to boost a resistance response to *Ophiostoma ulmi* (which causes Dutch elm disease) and, of the vaccinated trees, only 0.1 per cent have been infected by the disease.

Unlike human vaccines, which are specific to a single disease, a plant 'vaccine' improves all-round defences against a wide range of diseases. Researchers have recorded a buildup of antimicrobial proteins and defensive chemicals such as phenolics and terpenes. They also noticed that vaccinated plants produce thicker leaves that are less susceptible to pathogens, while conifers increase resin production. Dr Percival has carried out research into the use of soil amendments to reduce the severity of disease in trees, trialling substances such as chitin (the substance found in fungal cell walls and crustacean shells), phosphites and

Boosting defences

biochar, as well as the use of single tree-species mulches like willow, poplar and eucalyptus. All were found to be effective to some degree, particularly willow. As willow mulch decomposes, it releases salicylic acid, which stimulates roots and boosts the production of phenolic compounds that help the tree defend itself against disease. Phosphites have been sold for about 15 years as biostimulants. Chemically similar to phosphorus, they have been found to be effective against the fungus *Microdochium nivale*, a pathogen of lawn grasses. In other trials, it was discovered that phosphites are effective against apple scab and bleeding canker in horse chestnut trees.

I was concerned that the veteran oaks in my fields might succumb to acute oak disease, so I asked Dr Percival how I could boost their defences. He suggested a commercial product called Rigel-G that contains salicylic acid and garlic, and which gardeners can buy. It's used as a soil drench and applied under the canopy from the trunk to just beyond the drip line (the furthest edge of a tree's canopy from which water drips off the leaves). The salicylic acid induces a response in the roots. As always, it works better as a preventative treatment rather than waiting until signs of the disease are seen. The aim of the treatment is to create a hardier plant that will have improved vigour. The manufacturers claim that it also works against caterpillars, slugs and snails, vine weevil grubs, cabbage root fly, wireworm and leatherjackets, so I assume that may be the garlic element. Like all these types of treatments, you can't tell immediately if it is working as it's a long-term approach.

ROOT WARS

A lot goes on in the soil around plant roots. One phenomenon is called allelopathy, which is when one organism produces a compound that affects the growth or survival of another in the same ecosystem. Some of these interactions can be beneficial, others harmful. For example, walnuts, hairy vetch, peas, broad beans and buckwheat all have an ability to strongly suppress other plants.

Plant defences

There are interactions that affect pests and disease, too. For example, mustard deters leatherjackets and caliente mustard combats verticillium wilt in strawberries, while French marigolds can suppress nematodes (see 'Repelling pests', page 88).

Trees are good examples of allelopathic plants. They deter other trees from growing close by, so they can get all the nutrients and water they need from the soil. Species that are particularly good at this include the black walnut, eucalyptus, laurel, maple, pine, sumac and rhododendron. In the case of the black walnut (*Juglans nigra*), there are allelopathic chemicals in its leaves and nuts which fall to the ground and are released during decay, while its roots secrete juglone (an alleopathic compound) directly into the soil.

Given that the allelopathic chemicals survive decomposition and remain in the soil, gardeners need to be mindful of what they put in their compost bins, as there is the chance that the compost is then used to pot up plants where the residual chemicals could have an inhibitory effect on growth.

Biofumigation

We can take advantage of these allelopathic chemicals through a method called biofumigation. This a method of growing a cover crop, then chopping it up finely and turning it into the soil to release allelopathic compounds that adversely affect soil pests and weed seeds. I first became familiar with the idea of biofumigation when I grew a wild bird seed mix containing mustard as a strip along one side of a field. The following year, it was turned in and sown with wheat. It was a bumper year for wireworms (the larvae of click beetles) and much of the crop failed, but where there had been mustard the previous year, the wheat grew well.

Mustards contain a chemical called glucosinolate that makes them taste spicy, and this can work to deter pests, such as leatherjackets, wireworms and nematodes, in the soil. The caliente mustard variety is particularly high in glucosinolate and can be used to overcome fungal disease, such as verticillium wilt.

Boosting defences

Verticillium wilt is a soilborne fungus affecting a wide range of crops, trees, shrubs and ornamentals. It can wipe out whole crops of strawberries. Growing a cover crop of caliente mustard can release enough chemical into the soil to overcome the fungus and allow the crop to be grown again. Other plants with biofumigant properties include radish and forage sorghum, both relatives of the mustard. If you give biofumigants a try, remember that their effects are short-lived, so they usually only work against the current crop and do not necessarily provide any long-term protection.

Mustard has also been found to be effective against potato cyst nematode (PCN). This pathogenic nematode occurs around the world where potatoes are cultivated and is estimated to affect more than half the potato fields in England and Wales. It is a specialist pathogen and affects only potatoes, aubergines and tomatoes. It damages the roots of the crops, which causes poor water and nutrient uptake, and lower yields. Crop rotation is a good way of reducing the problem, but the nematode numbers decline very slowly. The only control is a nematicide, a pesticide that targets nematodes. But mustard could be an option. Grow mustard as a cover crop on the beds where you intend to grow potatoes. Just before flowering, cut it down and chop it into tiny pieces. Then turn the pieces into the soil where they will decay and release the chemicals. The key is to make sure there is plenty of moisture in the soil and that the mustard is mixed into the soil and not left as a mulch on the surface. Three weeks later, it will be safe to plant your potato tubers. This may not appeal to those who practise no-dig gardening, but it may be worth trialling if there are nematode issues in the soil. Two other plants to look out for are African nightshade (*Solanum scabrum*) and sticky nightshade (*Solanum sisymbriifolium*). These are being trialled by organic potato growers as trap crops for the nematodes. Research has shown that these two plants could reduce PCN.

CHAPTER 9

Barriers, lures, traps and sprays

There are some years in the garden when everything seems to conspire against you. A warm, wet spring can give that boost to the early pests, and then a long, dry summer puts your plants on the back foot while the pests thrive. You may have a healthy soil but, if the environment works against you, it will be an uphill battle to keep the pests and diseases at bay. There are, however, lots of non-chemical ways to tackle problem pests, through the use of physical barriers to prevent them gaining access to your crop as well as sticky traps and lures baited with pheromones. Also in your arsenal are fermented teas which can be used as a natural spray against pests and diseases.

PHYSICAL COVERS, BARRIERS AND TRAPS

Physical barriers, such as nets and covers, prevent pests gaining access to the crop. It is important that the barrier is in place before the seedlings emerge or immediately after transplanting. Once the crop is large enough to cope with damage or the pests have gone, the cover can be removed. Remember, though, that nets can reduce the amount of light reaching the crops and may weigh down on the crop unless supported by hoops. Netting covers vary in terms of their mesh size and weight. The finest nets, with a mesh of about 0.8–1mm, provide protection against tiny insects such as flea beetles, thrips, carrot and cabbage root flies. They are laid over crops and held down with pegs or stones. They can also protect against the elements, such as wind and hail, create a bit of shade, and give some protection from frost. Butterfly netting has a larger mesh, around 7–10mm (¼–½in), and this can be laid over the crop or supported on hoops or on fruit cages. But be warned: if the plants grow large and their leaves touch the mesh, the canny butterflies can lay their eggs through the gaps in the mesh. Personally, I like the softer butterfly netting rather than the stiffer variety which I find rips more easily. Bird netting has a larger mesh size, but don't go for anything more than 20 × 20mm (¾ × ¾in), since birds can get trapped.

Vertical fleece barriers are recommended to prevent carrot root fly from reaching your carrots. But research has recently found that the flies are actually much stronger fliers than originally thought and can actually fly over barriers, so the latest recommendation is to cover the whole crop.

While covers certainly can protect your crops from insect damage, there is a risk that they can inadvertently cause other problems. During the drought year of 2018, I covered my brassicas as usual with a fine butterfly net to protect against butterflies and pigeon damage. The plants survived the drought well, as the nets shaded the plants and held the moisture, but I had the worst plague of whitefly that I had experienced for years. The plants were smothered and every time I lifted the nets, I was met with clouds of whitefly. Why had the whitefly got out of control? I suspect it was because they had perfect growing conditions under the nets – it was moist and warm and there was also plenty of food and few predators, especially predatory flies and wasps that parasitise the eggs of the whitefly. So, I had no caterpillars but loads of whitefly and I'm not sure which was worse! Since then, I have always lifted the nets for a few days at a time, in order to give access to predators, and then checked the plants carefully for caterpillars before replacing the nets.

Barriers, lures, traps and sprays

Sticky traps

Yellow 'sticky paper' traps can be hung in the greenhouse or polytunnel to catch small flies, such as aphids, thrips and whitefly. They have been around for years and are basically cardboard or paper covered in a non-setting glue. They can be used to monitor the level of pests or hung in places where you know you have a problem. The traps are yellow because insects are attracted to the colour but, if you are controlling thrips, the advice is to use a blue colour. Sticky traps are placed about 20cm (8in) above the crop, as these insects do not fly long distances. They are easy to make, as you can just smear a piece of card with honey. But a word of warning: these sticky traps are not selective, so parasitic flies and wasps, and even bees, can get trapped.

Grease bands

Female winter moths are wingless and crawl up the trunks of trees to lay their eggs. The larvae hatch in spring and damage fruit buds. To stop their movement, grease bands can be wrapped around the trunks of fruit and ornamental trees, about 50–60cm (20–24in) above the ground, in mid-autumn and left until mid-spring. For bark that's quite knobby or has crevices, use a barrier of horticultural grease so that the moths can't creep underneath. If you have a good population of birds, there may be no need for the protection as the caterpillars form an important source of food for them in spring.

Pheromone traps

You can lure pests away from your plants by using pheromones. These are chemicals that are secreted by insects to influence the behaviour of other members of the same species. For example, female moths release a pheromone to attract the males. The pheromone wafts on the wind, often for quite some distance, and is detected by the male moths. Nowadays, synthetic pheromones can be used in a wide range of traps to attract specific pests. A pheromone dispenser releases the chemical at a specific rate and lures the male insects into the trap. Small insects can be trapped on sticky card, while funnel traps are used for larger insects; the pheromone draws them into the funnel where they slide down into the trap. Depending on the design, the traps can be hung from branches, suspended above crops or placed on the ground. Pheromone traps have to be in place before the insect becomes active, and they work by either trapping males, so there are fewer around to mate with the females, or simply by releasing a lot of pheromone that confuses the males so they can't find the females. Gardeners can buy pheromone traps for many pests, including box tree moth, apple moth, codling moth, tortrix moth, plum fruit moth, leek moth, pea moth, strawberry blossom weevil, horse chestnut leaf miner, raspberry beetle and thrips.

I use codling moth traps in my orchard. The caterpillars of the codling moth (*Cydia pomonella*) damage apples and pears, so this pheromone trap draws in the males. It doesn't have a huge range, so you need one trap for every five trees, hung about 1.5m (5ft) from the ground. I trap leek moths (*Acrolepiopsis assectella*), too. Leek moth caterpillars burrow into the stem of the leek, causing extensive damage. There are two generations, in spring and late summer. I suspend the trap above the plants, though I use it mainly as a monitor, as it alerts me to the fact that the moths are about and that I should cover the crop temporarily with a cover.

Are these traps worthwhile? It's difficult to tell, as this is a long-term strategy. If you use them every year, you should see a reduction in the levels of the pest and experience less damage.

HOMEMADE SPRAYS AND BREWS

Many gardeners have brews that they use against various pests and diseases, but these are not pest-specific and, if used too much or too widely, can harm beneficials and even damage your plants. I have suggested here how much to use, but there are no fixed rules and no guarantee that they will work. They don't keep for very long, so pour any leftovers on the compost heap. Also, it's wise to do a test spray to check the plant is not sensitive to the ingredients.

Barriers, lures, traps and sprays

Teas

These 'potions' are made by taking 1kg (2¼lb) of healthy leaf material, chopping it finely and then letting it stew in 10 litres (2½ gallons) of rainwater for between a few days and a few weeks, stirring daily or using an aerator. The resulting tea is filtered and sprayed over affected plants. These teas are like compost teas, but without the compost, and the residue in the filter can be put in the compost heap.

Nettle tea

Nettles are rich in nutrients such as nitrogen, phosphorus and calcium, along with silica, selenium and zinc, so can be used as a natural fertiliser as well as a pest deterrent. Nettle tea can be used against aphids and mites. Nettles take longer to break down, so leave the tea to brew for a couple of weeks.

Horsetail tea

This is one of my favourites, particularly since I have a patch of horsetails growing on the farm. It may be a pernicious, invasive weed in the garden, but it is useful for disease control because it's rich in silica. Silica helps to boost leaf cuticle health, which is critical to plant defences. Leave the stems for a couple of weeks in water that is stirred regularly or aerated. The tea is ready once the water turns a dark brown-black colour. Filter the tea and then spray it onto plants suffering from fungal diseases such as blight, mildew, rust, rots, scab, peach leaf curl and blackspot. You can even harvest horsetail and dry it, so you have a stash ready to use early in the growing season. If you can't find any horsetail, you can buy a commercial horsetail treatment instead.

Garlic sprays

The sulphur-rich compounds in garlic act as a repellent. Crush several garlic cloves and steep in 1 litre (34fl oz) of water for several days. If you want to make the spray more potent, add a few chopped fresh chillies. Strain the mixture, and then add more water to make up to 5 litres (1¼ gallons). It should smell quite strong! The spray is good as a repellent against sap-sucking insects and spider mites but can harm beneficials. Some gardeners use it to prevent powdery mildew. It can also work against rodents.

Essential oils

We value aromatic plants such as basil, lavender, mint and rosemary for their scent, but the volatiles also function as a deterrent or even a toxin to leaf-eating pests. I grow a lot of basil and mint each year, so it's easy to go out and harvest a large handful to make up a spray. Pour 5 litres (1¼ gallons) of boiling water over 500g (1lb 2oz) of basil or mint (peppermint is best). Cover and leave for a few hours, then strain and dilute 1:4 with more water. Spray the liquid over plants as a repellent. Alternatively, you can buy essential oil of mint and add 10 drops to 1 litre (34fl oz) of water. Initial research has found rosemary to be effective in controlling mildew on vines and it can be used in the same way as basil or mint.

There are three essential oils that you can buy: eugenol, geraniol and thymol. Eugenol is a phenol found especially in aromatic and spice plants, such as basil, clove, bay, cinnamon, marjoram and nutmeg. Geraniol is a terpene from aromatic plants and has been found to have insecticidal and repellent properties, as it affects the insect's nervous system. Thymol is derived from thyme and is used as an insect repellent and fungicide.

OILS, LIQUIDS AND SOAPS

There are various oils and liquids that you can use to smother the pests and make conditions less favourable for pathogens. However, although these are not harmful to the environment, they are not targeted and can kill beneficials as well as pests.

Vegetable/horticultural oil

These oils can be used against scale and aphids as they block their spiracles (breathing pores), causing them to suffocate. Blend 500ml

Barriers, lures, traps and sprays

Heat-treated seed

One of the reasons I look for organically certified seed for my garden is that I want to avoid treated seed. These are seeds with coatings or dustings of pesticide, which are there to protect the seed from soilborne organisms. The seeds are mostly treated against the diseases that affect the germinating seed and seedlings – for example, damping off by *Pythium*, *Rhizoctonia*, and leaf spots – but they can adversely affect other soil life, especially fungi. Fortunately, there are alternatives. Before fungicides came along, seeds would be subjected to a hot-water treatment. The temperature of the hot water was sufficiently high to kill the fungal pathogens on the seed coat and even inside it, but not hot enough to kill the seed.

Recently, a trial has been carried out by Innovative Farmers (a group of farmers and growers running on-farm trials), using hot water to treat chard seeds, with great success. The hot-water treatment is carried out immediately before sowing seed by placing the seeds in a muslin bag, suspending them in water at 50–52°C (122–126°F) for 10–20 minutes, then removing and drying them before they are sown.

Heat treatment works against diseases such as leaf spot, damping off, ring spot in chard, spinach and brassicas, *Fusarium* in squash, and bacterial blight and *Alternaria* in carrots. If you try this yourself, check the seed packet to make

Boosting defences

sure the seed hasn't already been heat-treated because you don't want to do it twice. While it is recommended for aubergines, brassicas, celery, chard, peppers, radish, spinach, tomatoes and turnips, it doesn't suit all seeds. It doesn't work for peas, beans, cucumbers, lettuce, beetroots or sweetcorn, for example, and it's not suitable for old seed as the heat treatment is likely to reduce germination. I have noticed that self-sown chard in my garden rarely suffers from leaf spot, but plants grown from purchased seed and bought-in seedlings have suffered. So, this either means I have clean seed or plenty of beneficial fungi in my soils to counter the pathogenic ones!

(17fl oz) of vegetable or horticultural oil with 125ml (4fl oz) organic liquid soap with no added chemicals, so it turns white. Then dilute the blended oil 1:10 with water and spray over pests. But remember that, being oily, the treatment will also block the plant's stomata and lenticels (which are used for gas exchange), so only use if absolutely necessary.

Milk

Milk can be diluted and used to protect against powdery mildew. Make up a 50:50 mix of milk and water, then spray over the leaves of plants such as courgette, squash and cucumber before the plants are showing signs of infection; it's best to start when there is some warm, humid weather forecast. It is uncertain why milk works but it's thought that when exposed to the sun, milk proteins have anti-fungal properties.

Barriers, lures, traps and sprays

Soap spray

Soap acts to disrupt the waxy waterproof covering of insects. For this, you need to use a pure soft soap or organic liquid soap, so there are no added chemicals. Add 5 tablespoons of soap to 4 litres (1 gallon) of boiling water. Cool and use as a spray over sap-sucking pests such as aphids. It's best to spray early or late in the day, rather than at midday, as the water will evaporate too quickly. Also, there are fewer beneficial insects around. Rinse the soap off the plants the next day.

PART 5 Managing water

CHAPTER 10

Be water-wise

With climate change making itself felt, now is the time to think about water and how you deal with extreme weather events. Will your garden or allotment cope with receiving a month's worth of rain in just a few hours or experiencing weeks of high temperatures and no rain? These new conditions cause problems for gardeners and allotmenteers wherever they are in the world, and they are becoming increasingly common.

Heatwaves and extended droughts can lead to hosepipe bans and, in some of the more remote areas, households reliant on private wells may run out of water completely and have to survive on tankered-in water supplies.

Water supply is already an issue on allotment sites. Some have mains water provided, while others rely on harvested water and water transported in by car. There may be mains water, but the number of taps may be limited and the use of hosepipe prohibited. Some sites put a limit on the amount of water that each allotment plot can use. And with water becoming an increasingly expensive commodity, this is going to be a big issue, as it is predicted that the UK will receive less rainfall in summer, combined with higher summer temperatures.

In the future, we need to be much more water-savvy in our gardens and allotments when we think about our everyday use of this vital resource.

WATER HARVESTING

Saving rainwater is one of the most important things to do in order to ensure you have plenty of water for the growing season. The obvious place to start is harvesting from roof surfaces. Many homes already have water butts with a capacity of between 200–300 litres, but there is potential to harvest so much more water from even the smallest of roofs. You can calculate just how much by multiplying annual rainfall (mm) by the roof surface area (m^2) to find the potential volume of rainwater in litres. You can easily find out your local rainfall online, but the average annual rainfall for the UK is 885mm. If we use a typical shed as an example, a 5m^2 area of roof will receive, on average, 4,425 litres of water a year. Experts estimate that you should be able to harvest around 75–80 per cent of the rain falling on the roof. That should be more than enough to get an allotment through a summer drought, providing you have enough storage. You can easily increase the area from which you harvest by building simple structures and adding roofing sheets, with guttering to direct the water into a series of water butts or containers. If you need more storage, look for second-hand rigid IBCs (intermediate Bulk Containers), which are relatively cheap, easily obtained and store one cubic metre of water (1,000 litres).

Dipping ponds

A simple dipping pond is a great way of storing larger quantities of water. As the name suggests, it's a pool of water into which gardeners dip their buckets. It's not populated with aquatic plants, but is purely for the purposes of storing a large quantity of water. In my walled garden, I rely on a dipping pond for my vegetable plot. It's fed from the roof of a nearby barn and holds around 3,000 litres of water. There is an overflow too, so when the pond is full, excess water drains into a nearby natural pond. It's easy to fit a diverter on a downpipe on the back of the house or other roofed structure and direct the water along a hosepipe into a dipping pond or even an underground tank for storage.

Be water-wise

How much water does an allotment need?

The National Allotment Society did a survey to find out how much water was used by allotmenteers and, as expected, the results varied a lot according to the location, local climate and soil type. On average, water use in the South West was around 1,800 litres per allotment per year. Being an average, it means that some allotments use far more, especially those with a greenhouse or polytunnel. But owning a greenhouse does offer a large roof space from which to harvest water!

Not surprisingly, the use of water is much less on those allotment sites where there is no mains water, for example, on my local allotment site. Here, the allotmenteers have no piped water so they have constructed a number of structures to support second-hand roofing panels and harvest a lot of water, which is stored in either IBCs or blue drums with a capacity of around 200 litres. So efficient is their water harvesting and soil management, that they had enough water to get through the droughts of 2018 and 2022.

It's a real time saver too, no more standing around while the bucket or watering can fills with water from a water butt or tap, it's just a quick dip. Because my buckets are filled so quickly, it also encourages me to drench the plants once a week rather than giving them a superficial sprinkle.

Bury a pot

A traditional method used by farmers in dry climates is to bury unglazed clay pots, called ollas, which are filled with water and then covered. I have been using old clay pots with blocked drainage holes to great effect in my greenhouse, in large containers and also in my veg beds. Water slowly permeates into the surrounding soil where it can be taken up by the roots. When the soil is dry, the water drains out quickly and the pots have to be replenished, but once the moisture levels are right for the plants, the period between refills increases. A similar set up can be achieved using old plastic bottles. A few tiny holes are pricked in the base and then the bottle partly buried in the soil near the plant. When filled, water slowly seeps out of the bottle into the soil. This method, combined with a good mulch, is probably the most frugal in regard to water usage, although it takes some time to set up.

Using grey water

Grey water is the waste water from sinks, washing machines, showers and baths (but not from the toilet). Using grey water from the washing machine is a good place to start as it is relatively clean and simple to access using a hosepipe. Another very simple, if less convenient, idea is to have a couple of buckets in the shower to catch water, especially the clean water that runs out before the hot water comes through.

Ideally grey water should be filtered as it goes into a collecting tank and it should not be stored for long periods. If you are going to use this water source, make sure you use eco-friendly and biodegradable soaps and detergents, and avoid those cleaning agents containing bleach and other harsh ingredients that could harm your plants or any insects on them. Don't use this water on edible crops, though and, where you do use it, make sure it goes onto the ground and doesn't splash the leaves of plants. Even eco-friendly detergents can damage plants and harm any insects hiding on them. Also, you do need to check local authority regulations regarding the use of grey water, as these do vary around the world.

Be savvy about your planting plans

Vegetables vary in their water requirements, for example lettuce, spinach and tomatoes are prone to water stress, whereas brassicas and leeks are more resilient. They have a longer growing season, so there is time to pick up again later in the year, especially during a mild autumn. It can therefore be useful to consider the different needs of the crops and how much water they are going to require. Those crops needing more water could be grown together, either in one bed, if appropriate, or in neighbouring beds of the garden or allotment, so they have the same watering regime. This way you can direct your water resources to the plants that need it most.

There are critical phases in the growth of plants that are more sensitive to water shortages than others. Beans and peas are susceptible to drought during the pod development phase while onions need water when they are bulking up their bulbs. Potatoes, with their large leaves and shallow roots, can be susceptible to drought, and insufficient water can lead to a substantial 80 per cent reduction in yields. There are two growth phases where having plenty of water is critical; the early stages of tuber formation and during flowering when the tubers are growing in size. Potato variety is important too, as some are more drought tolerant than others, for example, Cara, Marfona and Sarpo Mira. Not only is Sarpo Mira a blight-resistant variety, but it continues to grow well into autumn so can recover from a summer drought.

With summers likely getting warmer, it will be advantageous to get as much growing done early in the season, before the hot dry weather arrives, a pattern seen in the Mediterranean region. There may be fewer frosts in future and more growing days in spring and autumn, so we need to rethink our growing seasons and the crops we grow, especially through winter. I have been growing more

Managing water

autumn-sown crops, such as onions, chicory and broad beans. By getting the plants established over winter, they have a head start in spring, so they can be harvested before the heat of summer.

Similarly, an early sowing of peas can be harvested before a dry summer, whereas the later sowings may suffer in dry weather.

Shade is important too and it's often something that is overlooked. A shaded soil, for example, will not lose as much water by evaporation than one that is exposed to the sun. When planting your vegetable beds, think about casting shade over more drought-susceptible crops, for example by planting squash under runner beans and growing taller varieties of sweet corn and sunflower to cast shade over others. The three sisters combination of sweet corn, squash and beans (see page 97) shows how the crops benefit each other by casting shade and covering the soil as well as fixing nitrogen.

If you lack shade, you could put up arches to grow runner beans, peas and squashes or you could put shade cloth over crops to reduce the level of light. Many gardeners cover their brassicas, leeks and carrots with an insect mesh which has the added benefit of shading and reducing wind speed, so there is less evaporation from leaves on sunny days.

Perennial vegetables are a great drought-resilient option too, due to their established and often deep root systems and the relatively undisturbed soil around them, which leads to lower water demand.

Drought-resilient vegetables

It pays to study the drought resilience of the crops you grow, as some crops are better than others and there is even variation within a crop. For example, there are many varieties of lettuce to choose from, so opt for those that are more drought resilient, such as the loose leaf and oakleaf varieties. Drought-resilient crops include rocket, carrots, parsnips, beets and turnips, and you might consider growing French beans rather than runner beans, and Swiss chard rather than spinach. You could also consider trying some of the crops associated with warmer climates such as amaranth, chickpea, and okra (see box). It's a little experimental, especially as the weather is quite variable but as warmer summers become more frequent, it could be a wise option.

Some vegetables to try for warmer summers

Amaranth (*Amaranthus* spp.) Easy to grow, sown in late May, pick the young leaves and flower buds through summer.

Butter beans or lima beans (*Phaseolus lunatus*) A good alternative to runner beans, producing fat white seeds that can be eaten fresh or dried.

Chickpeas (*Cicer arietinum*) A sprawling plant with small pods, each containing one or two seeds for eating fresh or drying. The plant is frost-sensitive so the plants are best sown under cover and planted out after the risk of frost has passed.

Cowpeas or black-eyed peas (*Vigna unguiculata*) These beans (although called peas) are commonly grown in the southern USA, Africa and parts of Asia. These heat-tolerant legumes are easy to grow but need full sun and warmth. Sow under cover and plant out after the risk of frost has passed.

Edamame (soy) beans (*Glycine max*) A bush bean that is frost sensitive. Sow under cover and transplant outside in sunny position after the risk of frost. The pods develop at the same time and are harvested together. They must be cooked before eating. Don't bother to shell the beans as they simply slip out of the pods when cooked. If you want soybeans, leave the pods on the plant to ripen and turn brown.

New Zealand spinach (*Tetragonia tetragonioides*) A sprawling, low-growing and drought-tolerant plant that crops well in droughts, but it is frost sensitive. You may find that it's slow to get going, but it's worth the wait for the succulent leaves.

Okra (*Abelmoschus esculentus*) A sun-loving crop that came originally from Africa, but is now grown across the world. It is frost sensitive, so either sow under cover or outside after the risk of frost has passed (remember to soak the seeds before sowing). There are many different varieties, some of which have been selected to grow in the UK.

Tomatillo (*Physalis ixocarpa* and *Physalis philadelphica*) This Mexican native is a relative of the tomato and cape gooseberry, and is used to make sauces and salsas. It's a fast-growing, sprawling plant that benefits from staking. It is drought resilient and needs a long summer to mature. As the fruit ripens, the husk peels back to reveal a green- or violet-coloured fruit. There are

> some newer varieties available that are more suited to a temperate climate.
> **Yardlong beans** (*Vigna unguiculata* subsp. *sesquipedalis*) These tropical beans produce long, thin beans. They need heat so are best suited to growing under cover, but will grow well on a sheltered south-facing wall.

TOO MUCH WATER

In the years ahead, meteorologists are predicting more frequent and more violent storms and torrential rain. So, knowing how to deal with excess water will be paramount. Sadly, we will all have to get used to the sight of flooded lawns and waterlogged beds, but there are things we can do to make our growing spaces more resilient.

There are two key phrases that I can't emphasise enough. One is 'Slow it, spread it and sink it', which refers to the basic principles used by landscape architects to slow down the flow of surface water so that it spreads out and has a chance of soaking into the ground rather than running off into drains and water courses and creating floods. The other is 'Keep it permeable'. Our urban areas, and many of our parks and gardens have large areas of hard landscaping which means that water runs off these impermeable surfaces rather than soaking in. The more permeable our gardens, the less run off we get and the better it is for our communities.

When it comes to planning how to deal with too much rain, one of the best things you can do is to go out in a deluge and just watch. Observe what happens to water pouring off roofs and other surfaces, where it runs onto the ground and where it collects. You

Managing water

can use this information to help you adapt your garden or allotment. Here are some ideas:

Permeable surfaces. Avoid large areas of concrete and tarmac. Instead, use a permeable surface, such as gravel. If you need an area of hard standing or a garden patio, think about using permeable resin-bound gravel or permeable resin-based grout between paving slabs, so the water soaks into the ground rather than running off. Don't forget to ensure all your paths are permeable to avoid water using the paths as quick routes through the garden or plot.

Don't let surface water from your roof or other surfaces enter the drains. They should empty into a soakaway that carries the water away from the house to a lower-lying area from which it can slowly disperse.

Install French drains. These are small trenches that are backfilled with gravel, so they slow the flow and carry water away to somewhere where it can disperse.

Create rain gardens. A rain garden is a feature that manages surface water runoff. It's a shallow depression created on a gentle slope to temporarily hold surface runoff from roofs and paths. The excavated soil is put back into the depression and planted up so that, when it rains, the rain garden absorbs and stores water, allowing it to soak into the ground, rather than overwhelm the local drainage system. But do not think of a rain garden as a pond, since for much of the year, the rain garden is actually dry. However, the plants have to be able to cope with being flooded for a few hours at a time. As well as helping to reduce surface runoff, it can be planted

with pollinator-friendly plants to boost the biodiversity of the garden.
- **If you don't have space for a rain garden, consider a leaky rain barrel.** This is a gravel-filled barrel or container that is connected via a diverter to a downpipe so water collects and then drains slowly out of the bottom. It's acting as an attenuation vessel, as it intercepts rainwater and holds it temporarily, slowing the flow of water and reducing the risk of flooding.
- **Another simple yet effective idea to slow the flow of water is to empty your water butts.** A full water butt cannot slow water, so before a storm arrives, empty your water butts so they are ready to catch water (acting as attenuation vessels), helping to reduce the volume of water entering the local water courses.

Waterlogged soil

Waterlogged soils may become more common in the future when we experience more frequent periods of heavy rain. The key problem with waterlogging is that water fills up all the spaces around the soil particles, pushing out the air. This leads to a lack of oxygen for the plant roots and soil organisms alike, which means that they cannot respire aerobically. As a result, they soon run out of energy to fuel their metabolic processes. Soil life can cope if it's a bit of a flash flood and the water drains away quickly, but soil that remains waterlogged for days or longer means that microbes in the soil will start to die, as will the plant roots. The plants become stressed and may die too. In time, conditions can become quite anaerobic.

The first thing, and one of the most important things to remember, is not to walk on waterlogged soil or lawns – the weight will cause even more oxygen to be lost and the soil will be compacted which will just add to the waterlogging problems. This damage can take years to undo.

Check your plants and, if leaves start to turn yellow and there is some dieback, then they are suffering from waterlogging. Bizarrely, plants may be wilting even though the ground is waterlogged because their roots are dying. If it's a valuable plant, you may want to consider lifting it and temporarily placing it in a drier spot until conditions return to normal. Another point to remember is that plants that are stressed are more susceptible to disease, especially those caused by fungi and the damp conditions may favour pests such as slugs and snails.

Coping with a flood

Sadly, more gardens and allotments are likely to suffer from surface water flooding in the coming years. In the past, allotment sites were located in less valuable sites, particularly those not zoned for development, which often meant low-lying flood plains at risk of flooding.

If you know heavy rain is forecast and your garden or allotment is located in a high-risk flood area, there are a few things you can do to prepare. Firstly, empty those water butts. It is better to have empty water butts when the storm hits so that you can slow down the flow of water into water courses. If you have sandbags, use them to protect more valuable structures, such as greenhouses. Move as much as possible off the ground in your greenhouse and shed, especially any containers of fertiliser, pesticide and fuel, and remove valuable equipment, such as your tools and lawn mower, etc. Make sure electric equipment is unplugged and the water is turned off. It can be wise to place heavy objects on manhole covers too, so they don't lift and leave dangerous hazards under the water. If you have time, harvest your crops and stake any plants that are heavy with leaves or fruits.

Clearing up

It can be truly upsetting to see the mess that's left behind after a flood and its not just the damage from rainwater, but the contaminants that it carries, especially sewage. Don't rush outside to see

Be water-wise

what's happened but wait until the flood waters have gone and when you do venture out, make sure you wear protective clothing and waterproof boots, and keep your pets inside.

Rubbish and debris from a flooded garden is classed as controlled waste because there is a strong chance that it has been contaminated, so it needs to be disposed of carefully following advice from the local authority. Any woodchip, bark or compost etc. that has been covered by the flood will also be contaminated and there may be a layer of silt, deposited by the flood waters over all the surfaces. This should be removed wherever possible. What about your vegetables? They will be contaminated too, so don't ever eat any vegetables that have been flooded, including those vegetables that require cooking. In fact, it's not safe to use the soil for growing for at least a year after a flood event.

Flood water is a powerful force and can damage the foundations of buildings and retaining walls, so once the flood waters have subsided, check these structures, as well as any wooden arches and pergolas which may have been damaged.

The soil is going to be waterlogged for some time and, when it is in this state, it's easily damaged and susceptible to compaction, so don't walk on it and cause further damage. If you have standing water on some of your more valuable beds, it can help to dig some temporary ditches to let the water drain out.

Repairing the damage

A flash flood is unlikely to have done long-term soil damage, but it's very different if there has been standing water for days or weeks. The soil will have been well and truly waterlogged, it may have become anaerobic and the flood waters are likely to have carried away soluble nutrients, such as nitrates. Don't rush to do anything until the soil has had a chance to dry out and, if its winter, it can be better to simply leave the garden until spring.

Soil that has been flooded will benefit from the addition of organic matter, which will help to repair the soil structure and provide a source of nutrients. This can be dug in or if, like me, you

Managing water

are a no-dig gardener, cover the soil with a good mulch of organic matter. Another option is to sow a cover crop. It can also be beneficial to give a slow-release fertiliser to trees and shrubs, especially fruit trees and soft fruits in early spring.

My 10-point plan for a more water-resilient garden or allotment

1. Slow the flow of water and have permeable surfaces wherever possible.
2. Harvest rainwater to reduce your dependence on a piped water supply.
3. Make use of grey water to supplement rainwater in times of water shortage and hosepipe bans.
4. Water wisely. Water your plants either early or late in the day, so water has a chance to soak into the soil rather than in the heat of the day when much of the water will evaporate. Always give the plants a good soak so water permeates deep to the roots. This will encourage deeper rooting, leaving plants less susceptible to drought.
5. Build a healthy soil. Soils with a lot of organic matter will retain more moisture as well as have better drainage.

Be water-wise

6. Opt for no-dig beds (see page 22). A no-dig bed has a greater water-holding capacity than one that has been dug because there is better soil structure and improved soil aggregation that leads to more pores able to hold water. Long-term arable trials of no-till (no-dig) vs tilled in Ohio, USA, showed that no-till beds had higher water infiltration rates, so experienced less waterlogging.
7. Keep the network of fungal hyphae undisturbed by not digging. You definitely don't want to destroy the mycorrhizal fungi that are associated with many plants. We now know that in times of drought, mycorrhizal fungi can reduce drought stress in their host plants by increasing their own network of hyphae to find more water for their plant symbiont and they can even increase turgor within the plant, so reducing or delaying wilting. (Turgor being the pressure exerted by the fluid inside the plant cell against the cell wall, providing structural support)
8. Look after your earthworms. Those earthworm species that are vertical burrowers help to aerate the soil and improve drainage (see page 19). There is evidence that, in no-till arable soils with healthy earthworm populations, the rate at which water infiltrates the soil is as much as 6 times more than soils that are tilled.

Managing water

9. Mulch your soils to protect them from the sun and to help conserve water by reducing evaporation (see page 158). A deep mulch on a vegetable bed, whether inside a greenhouse, polytunnel or outside is essential, lowering evaporation by as much as 75 per cent compared with an unmulched bed, so reducing the need to water. As well as the more usual mulching materials, such as compost, leaf mould, woodchip and bark, you could consider a living mulch in the form of low-growing plants (see page 37).
10. Build raised beds and hügelkultur beds (see page 43) These beds improve your water resilience as they improve soil drainage. They are great for those drought-tolerant plants that don't like cold, wet soil in winter, and should there be any flooding, plant roots are above the level of flooding, unless it's very deep flood water.

A look to the future

The collective power of the gardening community as a crucial force for change should not be underestimated. There are around 24 million gardens in the UK, together with some 330,000 allotments, so boosting soil health and encouraging wildlife into these spaces would make a huge difference to the natural world. These small areas of wildlife habitat really hit above their weight, helping to provide better air quality, improved water retention, plus pollination services for insects as well as creating green corridors in urban areas.

Exchanging ideas and knowledge, collaborative efforts and community engagement all contribute to a collective resilience that go way beyond the individual garden or allotment. Shared experiences, educational initiatives and a sense of communal responsibility create a network of resilient gardens that collectively enhance the sustainability of the local environment.

CREATING RESILIENCE IN THE GARDEN

Throughout this book I have explored how we can improve the resilience of our growing spaces through the nurturing of our soils, the careful management of water and the creation of diverse spaces where wildlife can thrive alongside us.

Soil health is a cornerstone of garden resilience. Practices such as composting, mulching and the avoidance of digging all contribute to a thriving soil ecosystem. A robust soil structure not

A look to the future

only supports plant growth, but also enhances the garden's resilience by promoting nutrient cycling, water retention and the proliferation of beneficial microorganisms.

At the core of a resilient garden is biodiversity. By selecting a diverse array of plants, the gardener can create a mosaic of habitats that attract a range of beneficial organisms, from pollinators to predators. This biodiversity acts as a natural defence mechanism, as the intricate web of relationships between species helps mitigate the impact of pests, diseases and environmental fluctuations. The interdependence of these elements fosters a resilient balance, minimising the risk of widespread disruptions. Moreover, the adaptability of wildlife-rich gardens is a testament to their resilience. Unlike traditional manicured landscapes, biodiverse spaces tend to evolve, responding to changes in climate, seasons and the surrounding environment – an adaptability that helps to ensure a garden's endurance.

Water management plays a crucial role in achieving resilience too. Capturing rainwater as well as being able to cope with too much water are key to a water-wise approach, which not only conserves this precious resource but also enables the garden or allotment to withstand periods of scarcity, ensuring the survival of the plants during challenging conditions.

Resilient gardening emphasises organic and sustainable practices that avoid the use of chemicals and pesticides. By preventing the introduction of additional harmful substances to the surrounding air and ground, a resilient garden contributes to environmental health and the well-being of the broader ecosystem.

Awareness is yet another facet. Understanding local ecosystems, seasonal patterns and the needs of specific plant species empowers gardeners and allotmenteers to make informed decisions. The exchange of knowledge within a community of gardeners creates a knowledge bank that enhances the resilience of individual gardens and fosters a culture of sustainability.

A look to the future

WE NEED TO BE EXPERIMENTAL

Resilient gardens are adaptive, evolving with changing circumstances and gardeners employing these principles are open to learning and implementing new strategies. Whether it's experimenting with companion planting, integrating regenerative or permaculture principles or adopting new ideas around soil health, resilient gardeners embrace change and view challenges as opportunities for growth.

And there will be challenges. Nobody knows what's ahead in terms of the climate – it could be much warmer and drier, but it's just as likely that we don't get a Mediterranean climate, but one that is more topsy turvy, that varies from year to year, really challenging us in the garden. Going forwards, we need to be agile, adaptable and experimental, and to be prepared for failures.

As I listen to farmers and growers, I hear them say that no two years are the same in terms of challenges from the weather, one year the spring is warm and dry, the next it's cold. One grower said he planted more than 90 different types of fruits and vegetables, not because his market garden would suit them all, but he hoped that from this diverse selection, some would do well. And this is an important point that I have returned to several times in this book. We need to be more diverse in the ornamental plants and crops that we grow in our gardens and allotments, and this includes diversity in the crop varieties we choose in the hope that some will like the novel conditions. Looking ahead, it will be good to include new types of crops, maybe amaranth or chickpeas for example, as well as more perennial vegetables that tend to be more resilient to changes and challenges.

We need year-round diversity too. Our gardens are at their most diverse in terms of flowers in late spring and summer, but what about late winter? Many insects are active on a warm winter's day and they need food sources, so we need to provide for these early pollinators and predators. We can do this by making sure we provide plenty of food bribes from late autumn to early spring. It's not enough to have wonderful displays of daffodils and snowdrops. We need lots of different types of flowers and that includes weeds. Weeds are particularly important in the early and late months of the year, when they are often some of the few flowers available to insects, so a more relaxed style of gardening with a few weeds can really make a difference.

We can expect to see new pests, too. These may be completely new species that have migrated from further afield, such as species of longhorn beetles that are marching across Europe from Asia, defoliating trees as they go, or species that are expanding their range within a country, due to the changing climatic conditions. Milder winters, for example, enable them to survive when before they may have perished. It's important to be vigilant and be able to identify any new pest and its life stages, and to understand its life cycle. But once again, diversity in the garden and healthy soils will be key in keeping any new pest or disease in check. As I mentioned on page 17, mycorrhizal fungi have such an important role in plant health, and new research shows that treating soil with mycorrhizal fungi before sowing a crop can actually help the crop plants overcome fungal pathogens in the soil, avoiding the need to use fungicides.

NOW IS THE TIME TO PLANT TREES

Trees create valuable shade in the garden, but they take a long time to get established and they are more vulnerable to a late frost, summer drought or extreme cold during their first few years. With the climate becoming warmer and potentially drier, we cannot delay. As each year goes by, it will become more difficult to establish trees and they will require more aftercare in the form of watering. We need to act now to plant more trees for the future and, of course, we need to consider carefully which trees we plant as the climate will be very different in 30 years' time as they reach maturity.

A look to the future

PERSONAL RESILIENCE

Resilient gardens play a crucial role in enhancing our overall well-being too, having an impact on our physical, mental and emotional health. The physical activity of undertaking gardening tasks promotes cardiovascular health, flexibility and strength, while the growing and consuming of homegrown produce contributes to a nutrient-dense, healthy diet, which positively impacts our physical well-being.

The act of gardening, surrounded by nature, has been linked to stress reduction and relaxation, with the garden providing a peaceful and calming environment, allowing us to escape from the pressures of daily life and contributing to improved mental well-being.

Engaging with the ever-changing garden environment encourages personal growth, problem-solving, and a mindset of resilience that extends beyond the garden into daily life. Witnessing the growth and flourishing of plants provides a tangible sense of accomplishment, while the act of growing our own food gives a feeling of purpose and satisfaction. This sense of achievement contributes to our positive self-esteem and mental well-being.

In addition, resilient gardening involves a continuous process of learning, experimentation, and adaptation. Successful gardening involves keen observation, critical thinking and decision making –detecting early signs of plant stress, identifying potential pests or recognising soil problems all require a discerning eye. It encourages us to explore new ideas, take calculated risks and think outside the box when faced with challenges. These skills can transfer to other areas of life, improving our ability to analyse situations, identify problems and formulate effective solutions.

WHAT CAN THE GARDENING COMMUNITY ACHIEVE?

As a community of gardeners and allotmenteers, we have the chance to change the world around us for the better. By planting

A look to the future

for biodiversity, we can create habitats, albeit managed ones, in our own green spaces, which will attract insects, birds and more, and that effect is then multiplied by every garden or allotment that takes that approach.

Education plays a pivotal role too. By offering a tangible connection to the natural world, gardens, allotments and community spaces become outdoor classrooms. They cultivate an understanding of the delicate balance required for biodiversity, teaching gardeners and the wider community about the principles of sustainability and environmental stewardship. This knowledge empowers individuals to make informed decisions that enhance their garden's resilience and contribute to broader conservation efforts. And this is critical as we need to exert our influence with local planners: we need more green streets planted with trees for shade to create natural cooling areas as our world heats up, more green corridors to attract and support wildlife, and we all need to work together to slow the flow of water so that less surface water ends up going down the drains and creating localised flooding.

By looking after soil health, managing water and cultivating a wide and imaginative range of plants that supports wildlife, resilient gardening can become a force for positive change – a small, yet impactful step toward creating a more sustainable and resilient planet.

Index

Abbey Home Farm, 101
Abelmoschus esculentus, 160
African marigolds, 86, 91
African nightshade, 142
agroforestry, 100–101
air content, of soil, 10, 163
allelopathy, 41, 140–42
amaranth, 159
Amaranthaceae family, 18–19
Amblyseius mites, 127
amphibians, 73–74
anecic earthworms, 19
Anethum graveolens, 91
angelica, 82, 85
annual ryegrass, 33, 38
Anthocoris nemorum, 68, 79
Anthriscus cerefolium, 91
ants, 46, 63, 122–23
Aphidius wasps, 121, 124
Aphidoletes aphidomyza, 124–25
aphids
    and ants, 46
    biocontrol of, 118–19, 121, 124, 125
    climate change and, 60
    diseases carried by, 63
    predators of, 68, 73, 74, 79, 80, 84
    push-pull management of, 90
apple trees, 101–2, 114
archaea, 16
Armoracia rusticana, 91
Athelia rolfsii, 61

bacteria, 28, 56, 62, 129–30, 136
barrier crops, 90
barriers to pests, 143–46

basil, 88, 148
bats, 76–77, 80
beans, 40, 95, 97, 157, 158
beer traps, 52–53
beetle banks and buckets, 77
beetles, 67, 79, 128, 134, 172
beneficials. *See* natural predators
biennials, 112
Big Garden Birdwatch, 76
biochar, 27, 29
biofumigation, 141–42
biofungicides, 135, 138
biological control (biocontrol), 117–30
    with bacteria and fungi, 129–30
    benefits of using, 118–19
    concept of, 117–18
    with minute pirate bugs, 129
    with nematodes, 119–25
    with parasitic wasps, 121, 124–26, 129
    with predatory ladybirds, 128–29
    with predatory mites, 126–28
Biosave, 129–30
biostimulants, 137
biotrophs, 57, 63
birds, 74–76, 80, 106
black-eyed peas, 159
black walnut, 141
blight, 57, 61, 63, 108–9
blue tits, 74–76
bokashi compost, 30, 32, 98
bolting, 110, 111
boots, disease transmission on, 61–62
borage (*Borago officinali*), 82, 84–85
borers, 47
boundaries, 87, 105

Index

box blight, 60
brassicas
    compost for, 28
    continuous cropping of, 98, 99
    as cover crops, 33
    covering, 144, 158
    in crop rotations, 94
    lack of mycorrhiza in, 18–19
    mulching, 37, 38, 40
    pest control for, 63
    polyculture cultivation of, 95, 96
    saving seed from, 112
    trap cropping for, 89
    water resilience of, 157
breeding, 108, 112–14
broad beans, 99
brown composting materials, 24–28, 136
brown rots, 42, 58
buckwheat, 34, 38, 82, 84, 85
bug hotels, 68, 77–78, 105
bugs, beneficial, 68, 79
butter beans, 159
butterflies, 63, 71, 105–6, 144

cabbage root fly, 37, 63
Calendula officinale, 92
Calendula officinalis, 86
caliente mustard, 141–42
call duck, 54
cane toads, 118
carbon:nitrogen ratio, 25, 41–42
carbon sequestration, 5, 7, 11
carrot root fly, 63, 88, 91, 94, 144
carrots, 50–51, 88, 90, 91, 94, 144
caterpillars, 74, 75, 122–23, 132–35, 146
centipedes, 73
Cerevisane, 138
chafer grubs, 122–23
chalcid wasps, 71, 72
chamomile, 53
chard, 109, 149, 150
chemical defences, plant, 132–34
chervil, 91
chickpeas (*Cicer arietinum*), 159

clay, 10, 11, 15, 27
climate change, 60–61, 112–13, 139, 153, 171
codling moths, 73, 80, 122–23, 146
coffee grounds, 54
cold frames, 50
Coleman, Eliot, 94
collecting slugs and snails, 52
colour break, in flowers, 56
comfrey, 97, 117, 135
compacted soil, 13, 23, 31, 41, 163
companion planting, 4, 85, 88–89, 91–92
compost, 8, 22, 24–32, 36, 41, 50, 141
compost earthworms, 20
compost tea, 135–36
conifer-woodchip, 41
contamination, post-floods, 164–65
continuous cropping, 97–98
coriander (*Coriandrum sativum*), 82
corn, 40, 91, 97, 158
cottage gardens, 81
cotton, 108
courgettes, 37, 121
cover crops, 4, 21, 23, 32–35, 59, 94, 166
cowpeas, 159
crop mix, 51, 96
crop rotations, 59, 63, 93–95, 142
crumb structure, soil, 12, 13, 15
Cryptolaemus montrouzieri, 128–29
cucumber, 90, 117, 126, 127
cuffing crops, 23
Cylindrocladium buxicola, 60
cytokinin, 136, 137

dandelion, 86–87
debris, flood, 164
decomposition, 24, 25, 141
deflowering insects, 47
defoliators, 47
Deroceras reticulatum, 48
dill, 91
Dinocampus coccinellae, 71
dipping ponds, 154–56
disease
    climate change and, 60–61

Index

crop rotation to avoid, 99
garden management to prevent, 61–63
infected material, 39–41, 62–63
observation to control, 63–64
and population wheat, 113
stress and, 164
timing planting/sowing to avoid, 99
types of pathogens, 55–59
disease cycle, 55
disease resistance, 108–10, 115
disease tolerance, 109–10
diversity
biodiversity, 2, 65–66, 104–6, 170
crop diversity, 4, 21, 90, 92, 114
in resilient gardens, 4, 81, 170–72
slug control and, 49
of soil life, 7, 10–11
in vegetable plots, 92
within-species, 96
Dowding, Charles, 22, 98
drains, 162
drought, 85, 113, 144, 153–55, 158–61
ducks, 53–54

earthworms, 16, 19–20, 23, 39, 49, 167
earwigs, 73, 80
ectomycorrhizal mycorrhiza, 18
edamame, 159–60
Edmundson, Jill, 8
education, by gardening community, 174
effective microorganisms (EMs), 30
Empicoris vagabundus, 68
Encarsia formosa, 121, 124
endogeic earthworms, 19–20
endomycorrhizal mycorrhiza, 18
epigeic earthworms, 20
Erwinia species, 61
essential oils, 148
eugenol, 148
experimentation, 171–73
exudates, root, 16–17, 21

Fagopyrum esculentum. See buckwheat
feeding birds, 75
Feltiella acarisuga, 127

fennel, 82, 86
fermentation, 30, 32
fertilisers, 8, 9, 12, 50
F1 hybrids, 108, 110, 111
fire blight, 61
flea beetles, 89, 90
flies, 70, 83, 124–25, 128
flower bugs, 68, 79
flowers, 28, 82–86
Foeniculum vulgare, 82, 86
foliar sprays, 136
food, 5–6, 171–72
food waste, 30, 32
forest gardens, 97
forests, 22, 65, 93
French drains, 162
French marigolds, 86, 88–89, 91, 141
frogs, 73–74
frost-sensitive cover crops, 35
fungi
beneficial, 17–19, 23, 44, 167
biological control with, 129–30
compost rich in, 28, 103, 136
pathogenic, 42, 56–57, 61–62, 138
fungicides, 9, 132
fungus gnat, 124–25, 128
funnel traps, 146

gall-making insects, 47
gardening community, 169, 173–74
garlic, 90, 147–48
geraniol, 148
glomalin, 12, 14
glucosinolate, 141
Glycine max, 159–60
grafting, 114–15
grass, mowing, 50, 102, 103
grass clippings, 27, 28, 30, 36, 104–5
grass snakes, 74
gravel, 37
grease bands, 145
green composting materials, 24–28, 136
green-eyed wasp, 71
greenhouses, 121, 128, 164
grey field slug, 48

177

Index

grey water use, 156–57, 166
ground beetles, 67, 79
growing season, 157–58

habitats, creating, 2, 39, 75, 105, 169
hardening off plants, 50
hay, 28, 36
health, gardening and, 173
heat-tolerant plants, 159–61
heat-treated seed, 149–50
hedgehogs, 76
hedgerows, 28, 101, 103, 105
Helianthus species. *See* sunflower
Heracleum sphondylium, 84
Heterorhabditis bacteriophora, 122–23
Heterorhabditis megidis, 133–34
Hildebrandt, Urs, 26
hoeing, 23, 50
hogweed, 84
holly leaf blight, 60
Holmgren, David, 3
honeydew, 46, 58, 126
horseradish, 91
horsetail tea, 147
horticultural oil, 148, 150
hot-bin composting, 26–28, 63
hoverflies, 69, 80, 82–84
Howard, Albert, 5
hügelkultur beds, 44, 168
humus, 17, 24
Hymenoptera, 70
hyphae, 17, 18, 23
Hypoaspis mites, 128

Icerya purchasi, 118
Ingham, Elaine, 93
Innovative Farmers, 149
insect mesh, 158
insects, plant-eating, 36, 45–47
intercropping, 38, 89, 95, 96

Jerusalem artichoke, 86

keeled slugs, 48
kitchen gardens, 81–82

lacewings, 69, 79, 83, 125
ladybirds, 68, 71, 79, 83, 118, 125, 128–29
lawn, 8–9, 39, 104–5
leaf litter, 38–40, 49
leaf mould, 36, 38
leaky rain barrels, 163
leatherjackets, 122–25
leek moth caterpillars, 146
legumes, 34, 91, 93–94
lifting biennials, 112
lignin, 41
lima beans, 159
livestock, 22, 30
living mulch, 37–38, 87
loamy soil, 10, 13
Lobularia maritima, 82, 85
Lolium multiflorum, 33, 38
longhorn beetles, 172
low-hanging branches, 50
Lübke-Hildebrandt, Angelika, 26

maize biogestate, 36
manure, 11, 14, 22, 27, 28, 30
marigolds, 86, 91, 92
Maxmillian sunflower, 86
mealybugs, 128–29
medick (*Medicago littoralis*), 91
Mediterranean plants, 2, 43
metaldehyde, 54
Metaphycus helvolus, 126
microbes, 9, 131–32
mildews, 57, 58
milk, 150
millipedes, 73
mineralised wheat straw, 36
minerals, 10, 17
mint, 90, 133, 148
minute pirate bugs, 129
mites, predatory, 126–28
Mollinson, Bill, 3
monoculture, 90
monocultures, 92
morels, 18
mulching, 4, 35–43
    adding organic matter by, 11, 15

Index

after floods, 166
to attract natural predators, 78
for brassicas, 99
with cover crops, 35
with leaf litter, 38–40
with living mulch, 37–38, 87
in natural agriculture, 97
with no-work, deep mulch, 40
in orchards, 103
timing of, 51
typical materials for, 36–37
water resilience and, 168
with woodchip, 40–43
mustard, 89, 91, 141–42
mycorrhizal fungi, 17–18, 134, 167, 172

nasturtiums, 89–91
National Allotment Society, 155
national seed list, 111
natural agriculture, 97–98, 110
natural predators (beneficials), 4, 45, 65–80
  balance of pests and, 65–66
  biocontrol and, 117, 130
  encouraging, 36, 77–79, 103–4
  in orchards, 79–80
  and plant defences, 134–35
  planting to attract, 82–84
  of slugs, 50
  types of, 67–77
necrotrophs, 58, 62
nematodes, 19, 40, 53, 59, 89, 119–25, 133–34, 142
nest boxes, 75
nets, 143–44
nettles, 68, 91, 105–6, 147
newly-built homes, gardens of, 31–32
New Zealand spinach, 109–10, 160
no-dig gardening, 19, 22–23, 35, 98, 167
no-work, deep mulch, 40
nutrient density, 5–6
nutrient recycling, 17, 39

observation, 2–3, 63–64, 119, 161–62
oils, pest control, 148, 150

Okada, Mokichi, 97
okra, 160
olive scab, 60
ollas, 156
onion, 88, 90, 100, 111–12, 157
onion fly, 124–25
Oomycetes, 57
open-pollinated seeds, 110–11
Ophiostoma ulmi, 139
orchards, 62, 79–80, 101–4
ORC Wakelyns Population wheat, 113–14
organic farming, 5, 32, 35, 170
organic matter, 10, 11, 15, 16, 21, 38, 164–65
Orius species, 129

parasitic predators (parasitoids), 69–72
  flies, 70
  nematodes, 59
  wasps, 70–71, 79, 83, 84, 121, 124–26, 129, 133
parsnip (*Pastinaca sativa*), 85, 95, 112–14
PCN (potato cyst nematode), 142
peas, 95, 157, 158
Percival, Glynn, 139–40
perennial plants, 28, 40, 100, 110, 158
permaculture, 3–4
permeable surfaces, 161, 162, 166
personal resilience, 173
pest control, 64. *See also* biological control (biocontrol)
  covers, barriers and traps, 143–46
  and heat-treated seed, 149–50
  milk, 150
  mulch, 36
  observation and, 63–64
  oils, 148, 150
  push-pull, 90
  for slugs and snails, 51–55
  soap, 151
  teas and sprays, 146–48
  weeds and, 87
pesticides, 119, 132

Index

pests
    balance of natural predators and, 65–66
    climate change and, 59–60
    evolution of plants and, 2
    new, 172
    overreacting to, 66
    parasitic nematodes, 59
    plant-eating insects, 45–47
    predators of (*See* natural predators [beneficials])
    repelling, 49–50, 88–89
    slugs and snails, 49–55
    timing planting/sowing to avoid, 50–51, 99
pH, soil, 13–14, 54
phacelia (*Phacelia tanacetifolia*), 34, 82, 84
Phaseolus lunatus, 159
Phasmarhabditis hermaphrodita, 121, 124–25
pheromone traps, 145–46
Phillips, Michael, 103
phosphites, 139, 140
phyllosphere, 131–32
Physalis ixocarpa, 160–61
physical covers, 143–46
Phytophthora ilicis, 60
Phytoseiulus persimilis, 127
pine bark, 51–52
Pion, Martin, 18
plant breeding, 112–14
plant defences, 2, 131–42
    of allelopathic plants, 140–42
    and biofumigation, 141–42
    biofungicides to boost, 138
    biostimulants to boost, 137
    compost tea to enhance, 135–36
    foliar spray to enhance, 136
    in phyllosphere, 131–32
    supporting, 134–35
    "vaccines" to boost, 138–40
    volatiles as, 133–34
plant-eating insects, 45–47
planting, 81–106
    with agroforestry schemes, 100–101
    to attract natural predators, 82–84
    of barrier crops, 90
    of brassicas, 98–99
    companion, 88–89, 91–92
    in continuous cropping, 97–98
    with crop rotations, 93–95
    of easy-to-grow flowers, 82–86
    to increase crop diversity, 90, 92
    in no-dig gardens, 23
    in orchards, 101–4
    of perennial vegetables, 100
    in polyculture approach, 95–97
    to repel pests, 88–89
    and rewilding, 104–6
    of slug-deterrent species, 49–50
    timing of, 50–51, 99
    of trap crops, 89–90
    of trees, 2, 106, 172
    for water resilience, 157–58
    and weed management, 87
planting guilds, 96–97
plant pathogens, 55–59
plant selection, 107–15
    to attract natural predators, 82–84
    grafting, 114–15
    plant breeding, 112–14
    seed, 107–10
    seed saving, 110–12
    and slugs, 51
polyculture approach, 4, 95–97
polytunnel, 61, 62, 117, 119, 121
ponds, 106, 154–56
potato cyst nematode (PCN), 142
potatoes
    agroforestry with, 61
    biostimulants for, 137
    blight resistance in, 108–9
    continuous cropping of, 98
    in crop rotations, 94
    destroying infected, 63
    nematode treatment for, 125
    pathogenic fungi of, 57
    selecting varieties of, 51
    water resilience of, 157

Index

pot marigolds, 86, 92
pots, sterilising, 62
poultry, 22, 30
problem-solving skills, 173
Pseudomonas syringae, 129
pumpkin, 90
push-pull pest management system, 90
Pythium fungi, 57, 58

rabbits, 37, 43, 101–2
radish, 89, 91
rain, excess, 161–63
rain barrels, leaky, 163
rainforest, 2
rain gardens, 162–63
rainwater harvesting, 154–56, 166
raised beds, 43–44, 48, 53, 168
ramial chipped wood (RCW), 41–42
raspberries, 68, 86–87
Read, David, 18
red cabbage and lettuce, 109
red spider mites, 89, 117, 126–29, 134
regenerative gardening, 3, 32, 35
reptiles, 74
resilience, 1, 173
resilient gardens
    characteristics of, 4–5
    creating, 1–3, 169–70
    environmental impact of, 6
    experimentation in, 171–72
    and personal resilience, 173
    planting trees in, 172
    plant position in, 107
    soil health in, 5–6
rewilding, 2, 104–6
Rhaphanus sativus (radish), 89, 91
rhizosphere, 16–17, 44
RHS (Royal Horticulture Society), 51, 107
Rigel-G, 140
rock dust, 27
Rodale, J.I., 5
root exudates, 16–17, 21
rootstock, 114, 115
rot, leaf litter and, 39

Rothamsted Research, 12
rove beetles, 67, 128
rowan trees, 103–4
Royal Horticulture Society (RHS), 51, 107
Royal Society for the Protection of Birds (RSPB), 76
runner beans, 40, 158
rust fungi, 57, 58

salicylic acid, 138–40
sand, 10
sap suckers, 46–47, 56, 148, 151
Sarpo Mira potatoes, 157
Sarvari Trust, 109
scale insects, 80, 118, 124–26
sciarid fly, 124–25, 128
scion, grafting, 114, 115
sclerotium rot, 61
sea beet, 109
seaweed extract, 136
seeds, commercial, 108–9
seed saving, 4, 97–98, 108, 110–14, 113
seed selection, 107–10
semi-biotrophs, 57
semi-static composting, 24–26
shade, 158
shelter, for wildlife, 75, 101, 105
Shumei Natural Agriculture Farm, 98
silt, 10
slake test, 15
slow worms, 74
slug pellets, 54
slugs, 35, 36, 40, 43, 49–55, 120–21, 124–25
smut fungi, 57, 58
snails, 36, 43, 51–55
soakaways, 162
soap, 118–19, 151
social wasps, 71, 83
soft rot bacteria, 56
soil
    components of, 10–11
    flood damage to, 165–66
    importance of, 7
    plant defences beneath, 140–42

Index

soil (continued)
  theory of, 12
  waterlogged, 15, 163–66
Soil and Health Foundation, 5
soil health
  biostimulants to improve, 137
  current state of, 7–9
  and fertilisers, 9
  improving, 1
  in raised beds, 43–44
  resilience and, 169–70
  in resilient gardens, 5–6
  SOM as indicator of, 11
  tests of, 14–18
  and water resilience, 166
soil inhabitants, 7, 10–11, 17–20
soil organic matter (SOM), 11
soil pH, 13–14, 54
soil regeneration, 21–44
  biochar for, 29
  composting for, 24–32
  cover crops for, 32–35
  for gardens of newly built homes, 31–32
  mulching for, 35–43
  no-dig gardening for, 22–23
soil structure, 12–13, 14
Solanum scabrum, 142
Solanum sisymbriifolium, 142
SOM (soil organic matter), 11
southern blight, 61
South West England, 11, 155
sowing, timing of, 50–51, 99
soybeans, 159–60
spider mites, 89, 117, 126–29, 134, 148
spiders, 72–73, 79
spots, fungal, 58–59
squash, 38, 40, 90, 97, 158
Steinernema feltiae, 122–23
Steinernema kraussei, 124–25
sterilisation, of pots and containers, 62
sticky nightshade, 142
sticky traps, 145
Stout, Ruth, 40
straw, 28, 36, 40

strawberries, 135
structural defences, plant, 132
structures, flood damage to, 165
sulphur candles, 62
sunflower, 86, 97, 158
surface water flooding, 161, 164–66
sustainable gardening, 3, 170
sweet alyssum, 82, 85

Tachina grossa, 70
tachinid flies, 83
Tagetes erecta, 86, 91
Tagetes patula. *See* French marigolds
tansy (*Tanacetum vulgare*), 91, 92
Taraxacum officinale, 86–87
tar spot, 39–40
tea plant, 100
teas
  compost, 135–36
  pest control, 146–47
Tender and True parsnip, 112–13
Tetragonia tetragonioides, 109–10, 160
thread-legged bug, 68
'three sisters' vegetables, 97
thrips, 68, 69, 89, 91, 92, 127, 129, 145
thymol, 148
toads, 73
tobacco plants, 139
Tolhurst Organics, 101
tomatillo, 160–61
tomatoes, 38, 61, 63, 88–89, 92
tools, cleaning, 61–62
trap crops, 53, 89–90
traps, pest, 52–53, 143–46
trees
  agroforestry, 100–101
  allelopathic, 141
  bat-roosting sites in, 76
  climate change and, 60
  planting, 2, 106, 172
  'vaccinating,' 138–40
Trichogramma wasps, 129
Trifolium repens, 37–38, 94
Tropaeolum majus. *See* nasturtiums
tulips, 56

Index

Urtica dioica. See nettles
US Department of Agriculture (USDA), 6

'vaccines,' plant, 138–40
vegetable oil, 148, 150
vegetables, 5–6, 28, 40, 93–98, 157. *See also specific types*
Venturia oleaginea, 60
vertical spaces, 95, 106
verticillium wilt, 141–42
VESS (visual evaluation of soil structure), 14
Vigna unguiculata, 159
Vigna unguiculata sesquipedalis, 161
vine weevil, 63, 124–25
viral diseases, 56
visual evaluation of soil structure (VESS), 14
volatiles, 133–34, 148
voles, 43, 101–2

Wakelyns, 100–101, 113–14
wasps, 70–72
 chalcid, 71, 72
 parasitic, 70–71, 79, 83, 84, 121, 124–26, 129, 133
 social, 71, 83
water, for birds, 75
water butts, 154, 163, 164
water content, soil, 10
water infiltration test, 14–15
watering, timing of, 51, 166
waterlogged soil, 15, 163–66
water moulds, 57
water resilience, 4, 153–68, 170
 biochar and, 29
 as consideration during planting, 157–58
 coping with flooding, 164–66
 drought-resilient vegetables, 158–61
 grey water use, 156–57
 heat-tolerant plants, 159–61
 ollas and, 156
 planning for excess rain, 161–63
 rainwater harvesting, 154–56
 and soil structure, 13
 10-point plan to increase, 166–68
 waterlogged soil as stressor, 163–64
 and water use by allotments, 155
weeds, 8, 23, 24, 39, 41, 87
wheat, 57, 113
white clover, 37–38, 94
whitefly, 88, 121, 144
wildflowers, 32, 103, 105
willow, 41–43, 140
wilts, fungal, 58
windrow, composting, 26–28
winter moths, 145
wireworms, 141
woodchip mulch, 31–32, 40–43
wood mould boxes, 78, 105
wood piles, 49, 78
wool, 37, 51

Xylella fastidiosa, 60

yardlong beans, 161
yield, 107–8, 113

zombie ladybirds, 71

About the Author

Sally Morgan is an experienced organic, no-dig gardener, who loves to experiment and trial new crops. With a background in botany and ecology, she has always gardened naturally and is passionate about creating a growing space that it not just healthy and productive, but attractive to as wide a variety of plants and animals as possible. She has travelled widely, both in the UK and further afield, seeking out ideas and inspiration that can be incorporated into her own walled garden. Another source of inspiration are the books written by the pioneers of the organic movement – Lawrence Hills, Lady Eve Balfour and Sir Albert Howard and others – who understood the link between a healthy soil, healthy plants and healthy people.

Sally writes regularly for gardening and smallholding magazines and is currently the editor of *Organic Farming* magazine for the Soil Association. She is a member of the Garden Media Guild and her blog (livingononeacreorless.co.uk) was a finalist in the Garden Media Awards in 2019. She gives talks on climate change gardening and organic vegetable growing to groups across the country and runs courses on vegetable growing and self-sufficiency on her organic farm in Somerset, UK.

Her previous books include *Living on One Acre or Less* (Green Books, 2006) and *The Climate Change Garden* with Kim Stoddart (Cool Springs Press, 2022).

Twitter @Sally_Morgan
Instagram @the_organic_plot